TWO-DIMENSIONAL DNA TYPING
A parallel approach to genome analysis

ELLIS HORWOOD SERIES IN MOLECULAR BIOLOGY

Series Editor: Professor A. J. Turner, Department of Biochemistry, University of Leeds

Benne, R. **RNA EDITING: The Alteration of Protein Coding Sequences of RNA**
Eyzaguirre, J. **HUMAN BIOCHEMISTRY AND MOLECULAR BIOLOGY Vols 1-3**
Goddfellow, J & Moss, D. **COMPUTER MODELLING OF BIOMOLECULAR PROCESSES**
Parker, P. and Katan, M. **MOLECULAR BIOLOGY OF ONCOGENES AND CELL CONTROL MECHANISMS**
Sluyser, M. **MOLECULAR BIOLOGY OF CANCER GENES**
Tata, J. **HORMONAL SIGNALS: Cellular and Molecular Basis of Hormone Action and Communication**
Turner, A.J. **MOLECULAR AND CELL BIOLOGY OF MEMBRANE PROTEINS**
Uitterlinden, A. and Vijg, J. **TWO-DIMENSIONAL DNA TYPING: A Parallel Approach to Genome Analysis**

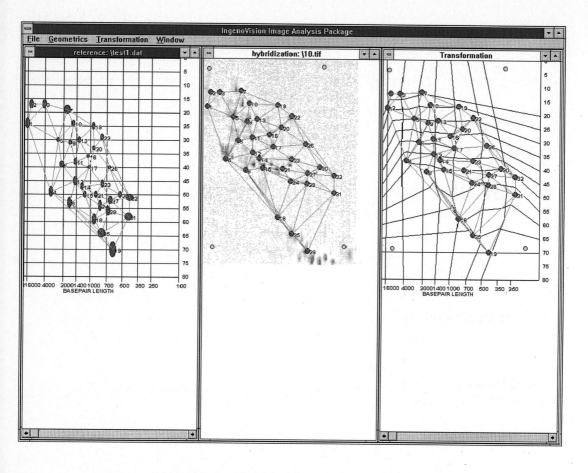

Frontispiece. The INGENYVision system. (a) Calibration step. A reference marker set (spot pattern on the left) is compared with the same marker set added as an internal standard to the genomic DNA samples to be analysed (pattern in the middle). The middle pattern shows the actual marker hybridization pattern of a particular gel with a genomic DNA separation pattern. The marker spots corresponding with those present in the reference marker set are indicated (numbers are arbitrary) and connected by lines to form triangles. Using triangle-transformation routines the reference pattern is transformed to match the internal standard of this particular gel. The resulting transformed reference pattern is shown on the right. The marker spots are indicated in green and the triangles in red. The yellow points in the corner are used for alignment of the marker pattern and the different core probe patterns obtained with a given gel. (See Fig. 3.12, p. 98.)

TWO-DIMENSIONAL
DNA TYPING
A parallel approach to genome analysis

André G. UITTERLINDEN, B.A., M.Sc., Ph.D.,

and

Jan VIJG, B.A., M.Sc., Ph.D.,
both of Medcand Ingeny
Leiden
The Netherlands

ELLIS HORWOOD
NEW YORK LONDON TORONTO SYDNEY TOKYO SINGAPORE

First published 1994 by
Ellis Horwood Limited
Market Cross House, Cooper Street
Chichester
West Sussex, PO19 1EB
A division of
Simon & Schuster International Group

Printed and bound in Great Britain by
Bookcraft, Midsomer Norton

Library of Congress Cataloging-in-Publication Data

Available from the publisher

British Library Cataloguing in Publication Data

A catalogue record for this book is available from the British Library

ISBN   0-13-791450-4 (hbk)

1 2 3 4 5   97 96 95 94

Table of contents

Preface

Application of the tools and approaches of molecular biology to basic problems of human disease has revolutionized biomedicine and will continue to bring fresh insight into important clinical problems. One of the most far-reaching applications of molecular biological methods has been in the area of hereditary or genetically acquired disorders. Here the techniques of DNA manipulation and analysis have led to the identification of an exponentially increasing number of genes, particular variants of which are responsible for one of the many thousands of disorders. On identification of the gene(s) involved in these disorders, the use of selected molecular probes opens up a new area of prenatal diagnosis, informed genetic counselling and genetic epidemiology. Provided that generally applicable 'gene delivery' systems will become available, gene therapeutics can provide a rational and effective alternative to the symptom-based drug treatment of these diseases.

The consideration that large-scale gene identification will have almost immediate applications in diagnosis and therapy explains the increasing public, governmental and corporate interest in the mapping and sequencing of the entire human genome. In addition, there is a strong interest from the agro sector and veterinary medicine in the identification of economically important traits. This is exemplified by the initiation of comparable genome projects on other species.

Human diseases have been shown to have genetic factors that influence their expression. It can therefore be expected that considerable emphasis will be placed on the development of techniques to identify rapidly the gene variants relevant for a specific disorder among the estimated 100 000 genes embedded in about 3 billion basepairs of human DNA. Two-dimensional DNA typing is a technique that allows genomes to be comparatively analysed, varying from small microorganismal genomes to the much larger genomes of higher animals. In combination with recently emerged systems for rapid data analysis and organization of information, the method should be useful to link established disease phenotypes to particular DNA sequences, thereby allowing a steady flow of genes to enter programs for DNA-based diagnosis and therapy.

This volume is the first that specifically addresses the principles and applications of 2D DNA typing, providing experimental set-ups, protocols and guidelines for equipment and reagents. The description of the 2D DNA typing technique is placed against the background of emerging techniques for analysing DNA and their application in genome mapping projects. Hopefully this book will contribute to the field of genome analysis.

We are grateful to our colleagues Drs W.J.F. de Leeuw, A.M. Verwest, and Y. Wu (INGENY BV, Leiden, The Netherlands), I. Meulenbelt and P.E. Slagboom (IVVO-TNO, Leiden, The Netherlands), A.C. Molijn (Erasmus University, Rotterdam, The Netherlands), B. Morolli (Leiden University, The Netherlands), G. Muyzer (Max Planck Institute München, Germany), and G.J.J.M. Trommelen (GenePharming BV, Leiden, The Netherlands), for sharing their results with us and especially E. Mullaart (INGENY BV, Leiden, The Netherlands) and G.J. te Meerman (Groningen University, The Netherlands) for their major contributions. We would also like to thank the following organizations and companies for their trust and financial support in development of the 2D DNA typing technology: the Netherlands Organization for Applied Scientific Research TNO (The Hague, The Netherlands), INGENY BV (Leiden, The Netherlands), Medscand AB (Malmö, Sweden), Toyobo Co., Ltd (Osaka, Japan), Gist Brocades NV (Delft, The Netherlands), Holland Genetics V.O.F. (Arnhem, The Netherlands). The Dutch Ministry of Economic Affairs (SENTER, The Hague, The Netherlands), and the European Community Human Genome Programme. We are indebted to our collaborators in the use of 2D DNA typing Professor Dr. A.-L. Børresen (Radium Hospitalet, Oslo, Norway), Dr T.A. Kruse (Aarhus University, Denmark), Dr H. Lehrach (Imperial Cancer Research Fund, London, England), and Dr P. Nürnberg (Humboldt Universität, Berlin, Germany) for their valuable input. We also gratefully acknowledge the enthusiastic support from the very beginning from Professor N.C. Seeman (New York University, New York, USA), Professor P.H.M. Lohman (Leiden University, The Netherlands), and Professor C.H.C.M. Buys (Groningen University, The Netherlands). Finally, we are greatly indebted to Drs Stuart G. Fischer (Columbia University, New York, USA) and Leonard S. Lerman (Massachusetts Institute of Technology, Cambridge, USA) for introducing us to the field of denaturing gradient gel electrophoresis.

December 1993

André G. Uitterlinden
Jan Vijg

List of abbreviations

APRT	adenine phosphoribosyl transferase
APS	ammonium persulfate
BSA	bovine serum albumin
CCM	chemical cleavage of mismatched basepairs
CDGE	constant denaturant gel electrophoresis
CEPH	Centre d'étude polymorphisme humain
CF	cystic fibrosis
CFTR	cystic fibrosis transmembrane regulator
CGH	comparative genome hybridization
cM	centimorgan
COL1A1	type I procollagen
CTAB	cetyl trimethyl ammonium bromide
DM	dystrophy myotonia
DMA	Duchenne muscular dystrophy
ds	double stranded
DNA	desoxyribonucleic acid
DGGE	denaturing gradient gel electrophoresis
DTT	dithiotreitol
EDTA	ethylenediamine tetraacetic acid
FIGE	field inversion gel electrophoresis
GMS	genomic mismatch scanning
HD	Huntington's disease
HERV-A	human endogenous retrovirus type A
Hex A	β-hexaminidase A
HLA	human leukocyte antigen
HMD	high melting domain
HPRT	hypoxanthine phosphoribosyl transferase
HTF	Hpa II tiny fragments
IAP	intracisternal A particles
IGF1R	insulin-like growth factor 1 receptor
irPCR	inter-repeat PCR

kb	kilobasepairs
LDL	low density lipoprotein
LMD	low melting domain
LOD	logarithm of the odds ratio
LOH	loss of heterozygosity
LTR	long terminal repeat
Mbp	megabasepairs
MCC	mutated in colorectal cancer
MHC	major histocompatibility complex
OTC	ornithine transcarbamoylase
PAA	polyacrylamide
PAH	phenylalanine hydroxylase
PBS	phosphate buffered saline
PCR	polymerase chain reaction
PFGE	pulsed field gel electrophoresis
QTL	quantitative trait locus
RAPD	random amplified polymorphic DNA
RDA	representational difference analysis
RNA	ribonucleic acid
RFLP	restriction fragment length polymorphism
RLGS	restriction landmark genome scanning
RSP	restriction site polymorphism
RTVL-H	retrovirus-like histidine tRNA (HERV-H)
SBMA	spinal and bulbar muscular atrophy
SDS	sodium dodecyl sulfate
ss	single stranded
SSC	saline sodium citrate
SSCA	single strand conformation analysis
SSCP	single strand conformation polymorphism
STS	sequence tagged site
TEMED	N, N, N', N'-tetramethylethylenediamine
TGGE	temperature gradient gel electrophoresis
TGF-α	transforming growth factor α
UF	urea/formamide
UV	ultraviolet
VNTR	variable number of tandem repeats
YAC	yeast artificial chromosome

1

Two-dimensional DNA typing: background and technology

1.1 ANALYSIS OF DNA: GENERAL ASPECTS

After the gradual recognition of DNA as the key molecule of life on earth, the last three decades have seen a tremendous expansion of DNA research. Especially the field of genetics has greatly benefited from the surge in new methodologies for investigating the transmission, mutation and action of genes in a variety of species. One of the major aims of genetics is to study variation. Genetic variation is basic to speciation and can underlie differences between developmental stages, including ageing, and between a normal and a pathological state.

To unravel the unique genetic basis of the many appearances in which life on earth manifests itself there is a growing need for techniques to characterize DNA molecules physically and to distinguish different DNA molecules from each other. Even the little that we already know about the DNA molecules that genetically define the earth's different life forms exhibits a bewildering variety of complexity. Differences in organization, structure and size of the genetic material in various organisms have become apparent and are often directly reflective of form and function. Indeed, even subtle differences in the DNA of the genome of higher organisms, such as that of humans, can have dramatic consequences for health and disease. Hence it is not surprising that DNA research has been fuelled by the development of powerful methods of analysis. These allow not only the characterization of a particular DNA molecule itself but also the study of species specificity and the comparison of near-similar (homologous) DNA molecules, e.g. of different, sometimes closely related species or individuals of the same species.

Early methods of analysing DNA molecules were focused on very generic characteristics of nucleic acids, such as chemical composition and atomic structure as measured, for example, by X-ray diffraction of DNA crystals. Later, more refined physical and chemical methods, such as those based on the temperature dependence of certain structural characteristics of the DNA molecule, revealed information on its composition in various organisms. This demonstrated a uniformity of structure and at the same time an extreme diversity in the composition of DNA molecules from different organisms. Of the differences the most striking involve the size of DNA molecules and their basepair sequence. In general, the genomes of lower organisms are smaller than those of higher organisms and are also much more compact. That is, there is a more efficient use of the coding capacity, in terms of the ratio of protein-coding vs non-coding basepairs, in lower organisms where genes are tightly packed with very little spacing between them. In Table 1.1 an overview of the genome sizes of various organisms is presented.

Table 1.1. Genome sizes of various organisms

Organism	Common name	Genome size (bp)
Viruses		
SV40		5 243
Bacteriophage φX174		5 386
Adenovirus		35 937
Bacteriophage lambda		48 502
Bacteriophage T4		70 000
Bovine herpes virus type I		138 000
Chlorella		330 000
Prokaryotes		
Mycoplasma capricolum		0.7×10^6
Haemophilus influenzae		1.9×10^6
Staphylococcus aureus		2.9×10^6
Clostridium perfringens		3.1×10^6
Bacillus amyloliquefaciens		4.0×10^6
Escherichia coli K		4.4×10^6
Escherichia coli B		5.0×10^6
Lower eukaryotes		
Amoeba dubia		6.7×10^{11}
Streptomyces albus	Yeast	1.0×10^7
Saccharomyces cerevisiae	Yeast	1.5×10^7
Aspergillus niger	Fungus	1.5×10^7
Aspergillus nidulans	Fungus	3.1×10^7
Neurospora crassa	Mould	4.7×10^7
Dictyostelium discoideum	Slime mould	5.4×10^7
Tetrahymena thermophilae		20×10^8
Insects		
Drosophila melanogaster	Fruitfly	1.8×10^8
Nematodes		
Caenorhabditis elegans	Roundworm	1.0×10^8
Vertebrates		
Tetraodon fluviatilus	Pufferfish	2.6×10^8
Scombridae	Tuna	6.0×10^8
Gallus domesticus	Chicken	1.2×10^9
Vespertillio	Bat	1.5×10^9
Rattus norvegicus	Rat	2.3×10^9
Mus musculus	Mouse	2.7×10^9
Xenopus laevis	Frog	3.1×10^9
Bos taurus	Cow	3.2×10^9
Homo sapiens	Man	3.4×10^9

(continues)

Table 1.1. *(continued)*

Organism	Common name	Genome size (bp)
Plants		
Arabidopsis thaliana		7.0×10^7
Gossypion	Cotton	5.2×10^8
Lycopersicum peruvianum	Tomato	3.0×10^9
Zea mays	Maize	3.9×10^9
Nicotiania tabacum	Tobacco	4.8×10^9
Allium cepa	Onion	1.8×10^{10}
Lilium formosanum	Lily	3.6×10^{10}
Mitochondria		
Homo sapiens	Man	16 569
Drosophila melanogaster	Fruit fly	19 500
Aspergillus nidulans	Fungus	33 250
Saccharomyces cerevisiae	Yeast	74 000–85 000

As a consequence of their small size and their relative simplicity in organization the genomes of lower organisms, such as viruses, were among the first to be characterized by recombinant DNA techniques. Recombinant DNA techniques allow, in contrast to earlier methods of DNA analysis, the information content of a DNA molecule to be probed directly. That is, by employing a combination of specifically acting DNA-modifying enzymes and analytical techniques such as electrophoretic separation and hybridization, insight can be obtained into both the size and sequence content of the genomes under study.

The first comprehensive genome analysis in this respect was the sequence determination of the complete human mitochondrial genome by Sanger and colleagues in 1981 (Anderson, S. *et al.,* 1981). Larger genomes, such as that of humans, pose a more formidable problem. Indeed, the human genome of more than 3×10^9 basepairs (bp) in size is so large as compared to the human mitochondrial genome of 16 569 bp that a simple Eco RI restriction enzyme digestion of human genomic DNA results in a million different fragments rather than 3 for the mitochondrial genome.

This extreme level of complexity of the higher animal and plant genomes has led researchers to focus their approaches to 'meaningful' areas in the genomes of interest, that is, the protein-coding sequences. According to this concept genes can be isolated by starting from proteins and subsequently identify the mRNA and the corresponding gene in the genome. This approach has been successful in identifying genes in a few hundred cases, including the identification of disease genes, but necessarily awaits the characterization of more proteins. Since a typical human cell can contain 10 000–50 000 different proteins, which are difficult to screen and propagate on an individual basis, this approach is not trivial.

A more rational approach to identify genes defining a particular trait is to associate the occurrence of that trait in individuals with the presence of a variant DNA fragment. The best example in this respect is a method now known as positional cloning through genetic linkage analysis. In this approach pedigrees of individuals are analysed for the statistically significant co-segregation of variants (alleles) of DNA markers together with a particular phenotype, such as disease in humans or an economically advantageous trait in animals or plants. Co-segregation is then explained by the close physical proximity on the chromosome of the marker locus and the gene responsible for the particular phenotypical characteristic. This will be discussed in more detail in Chapters 2 and 4. Central, however, to the success of this approach is the possibility of probing the genome for variation at DNA marker loci. Methods are therefore needed to detect in a short time as much of the variation present in the genome as possible. Naturally, the larger the genome of interest the more sites have to be scanned for variation. In general, in higher organisms one marker locus has to be analysed once every one to ten million basepairs to establish statistically significant linkage of the marker with the particular phenotypic characteristic in pedigree analysis. For larger genomes, therefore, hundreds to thousands of markers have to be analysed in each individual of the pedigree(s).

Another example in which rapid genomic analysis for DNA sequence variation would be important, is the analysis of tumours for prognostic markers, i.e. that predict the grade of malignancy. A comparative analysis of the tumour and nor-mal host genome for genomic differences will identify particular chromosomal regions containing genes involved in particular stages of tumour initiation and progression (see Chapter 4).

In principle, Southern blot hybridization analysis (Southern, 1975) has been the method of choice for the kind of studies referred to above. In this method the electrophoretic separation patterns of restriction enzyme digests of the DNA molecules of interest are transferred onto a nitrocellulose or nylon (DNA binding) membrane. By hybridization analysis using a radioactively labelled probe, specific for a particular genomic sequence, homologous fragments within the separation pattern are visualized after autoradiography (schematically depicted in Fig. 1.1).

Nowadays other (mostly electrophoretic) systems for directly measuring DNA sequence variation have been developed. Such methods, which will be reviewed below, offer the possibility of scanning the genome for DNA sequence variation by measuring different sites serially. By contrast, two-dimensional DNA typing, which is the subject of this volume, combines two independent electrophoretic separation criteria, including denaturing gradient gel electrophoresis (DGGE), in a two-dimensional analysis of DNA fragments. Thereby, hundreds to thousands of fragments can be resolved allowing genome scanning for DNA sequence varia-tion at a great number of sites simultaneously. Such a parallel processing approach is efficient and greatly increases the scope of genome analysis studies.

In the next sections the possibilities for measuring DNA sequence variations will be listed and discussed. In view of the use of DGGE in 2D DNA typing this electrophoretic separation technique will be discussed in some detail.

When analysing DNA sequence variation a fundamental distinction should be made between the analysis of known vs unknown DNA sequence variations. Once a sequence variant is known, one can apply a number of recently developed methods for its rapid and accurate detection. Some of these methods are well suited for the simultaneous analysis of large numbers of samples for a particular mutation and are referred to as 'mutation screening' methods. They are, however, generally not capable of assaying a given DNA region for unknown mutations. These techniques, therefore, fall beyond the scope of the present volume and will not be further discussed.

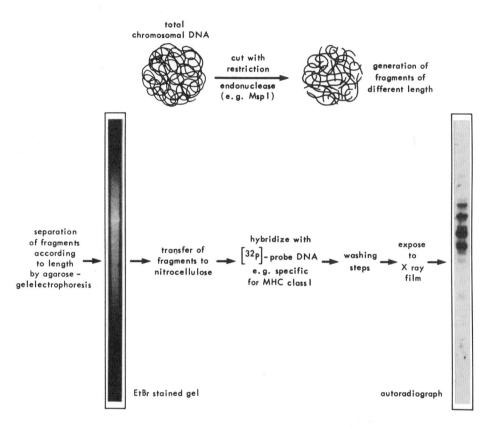

Fig. 1.1. Schematic representation of Southern blot hybridization analysis. In this particular experiment a cDNA probe for an HLA (the human major histocompatibility complex, MHC) class I antigen was used. Since the HLA complex on human chromosome 6 is a multigene complex the probe detects multiple sequences, i.e. restriction fragments derived from class I genes.

When within a given DNA region new, previously undefined, DNA sequence variations can be expected, techniques have to be employed which allow a considerable number of basepairs to be simultaneously analysed or 'scanned' for variation. Table 1.2 lists the currently available DNA scanning methods with their mode of action and their sensitivity in terms of the number of basepairs that can accurately be scanned for variation. These methods will be discussed below in terms of their suitability for serial genome scanning and, more specifically, their applicability as a second dimension criterion in a two-dimensional separation format.

Table 1.2. Efficiencies of 'mutation scanning' methods

Method	Number of basepairs scanned per lane	Size range test fragment (bp)	Efficiency[a]
DNA sequencing	400	10–400	100%
RFLP	10–20	500–10000	≤1%
RNase A	250–500	100–1000	≤70%
Chemical cleavage	50–1000	50–1000	≤99%[b]
DGGE	250–500	100–1000	
Unclamped:			
Homoduplex			≤40%
Heteroduplex			≤50%
Clamped:			
Homoduplex			≤75%
Heteroduplex			≤99%
SSCP			
DNA SSCP	150	50–600	≤95%[c]
RNA SSCP	150	50–600	≤75%[b]
ddF	300	300	≤99%
Heteroduplex mobility	50–1000	100–1000	[d]

[a] Percentage of all possible base changes detectable in a given fragment.
[b] Data based on the analysis of a limited number of sequences.
[c] Size dependent (Sheffield *et al.*, 1993).
[d] No data available yet.

1.2 MEASURING DNA SEQUENCE VARIATION

1.2.1 DNA sequencing

The ultimate identification of DNA sequence variation is the comparative determination of the basepair sequence of a wild-type molecule and its variants. In essence, DNA sequencing involves the generation of end-labelled fragments of different size which are electrophoretically separated in a polyacrylamide gel. Since 1977 two methods have been available and refined to sequence DNA fragments of up to 400 bp. They are based on nucleotide-specific chemical cleavage (hydrazine for T and C, piperidine for A and G (Maxam and Gilbert, 1977; Gilbert, 1981)) or the use of chain-terminating inhibitors of DNA polymerases (dideoxynucleotides (Sanger *et al.*, 1977)). In the chemical sequencing protocol the differently sized fragments find their origin in reaction conditions that permit on average only one break per molecule at any of the nucleotides for which the chemical is specific. In the enzymatic sequencing protocol fragments of different size arise as a result of particular ratios of natural dNTPs to any of the chain-terminating analogues which halt the polymerase at different positions along the molecule. In both protocols the rungs of the ladder of bands which arise after size separation of the fragments in polyacrylamide gel electrophoresis indicate the position of particular nucleotides in the DNA molecule of interest.

Almost all DNA sequencing reactions nowadays are done according to the enzymatic protocol of Sanger, mostly because of its simplicity and accuracy. The protocol has undergone considerable refinement by using modified polymerases such as T7 polymerase or Sequenase (Tabor and Richardson, 1987) and by using fluorescent dideoxynucleotides. The latter are used for non-radioactive DNA sequencing where the detection is based on laser excitation using automated instrumentation (Edwards *et al.,* 1990). Further improvements include the use of the polymerase chain reaction (PCR; Mullis and Faloona, 1987) for amplification of the fragment of interest, making cloning obsolete. These direct DNA sequencing protocols for determining DNA sequence variants (Yandell and Dryja, 1989) have been applied in, for example, the analysis of Lesch–Nyhan mutations in the HPRT gene (Gibbs *et al.,* 1989a), and the characterization of mutations in the factor VIII and IX gene (Higuchi *et al.,* 1990; Koeberl *et al.,* 1990).

DNA sequencing provides the ultimate tool in the detection of DNA sequence variation. However, for measuring variation in large DNA regions it is necessary first to subdivide the region in smaller sections of about 400 bp. Hence, large numbers of sequencing reactions must be performed on each sample. To become applicable for scanning or screening purposes, DNA sequencing requires dramatic improvements in speed. Although more rapid methods for sequencing are at present under development (Mathies and Huang, 1992), high speed at low costs can, most likely, only be achieved by methods other than electrophoretic separation. Such methods, for example parallel sequencing by hybridization (Drmanac *et al.,* 1989; Strezoska *et al.,* 1991; Dramanac *et al.,* 1993) and laser-based single-molecule detection in flowing sample streams (Davis *et al.,* 1991) are currently still under development but hold promise as fast and cheap sequencing methodologies.

From a conceptual point of view DNA sequencing for assaying interindividual variation is unattractive since it provides much unnecessary information; by far the largest part of the sequence to be comparatively analysed is identical. It is therefore advantageous to have methods available that only and uniquely identify deviations from a particular wild-type sequence.

1.2.2 Restriction enzyme digestion: RFLP analysis

The very start of recombinant DNA research was the discovery of enzymes capable of recognizing and cutting a particular short sequence of basepairs (usually 4–8 basepairs long) in a double-stranded DNA molecule, thereby generating staggered double strand breaks in the DNA molecule. The specificity of these enzymes with respect to the particular basepair sequence recognized appeared to be determined by the bacterial species from which they were isolated. In the bacteria themselves these enzymes serve as part of a host-defence (host-restriction) system to safeguard the bacteria from intruding non-self DNA molecules from other strains or bacteriophages.

Restriction enzymes were subsequently used in a comparative way, by digesting near-similar DNA molecules in parallel and analysing the different restriction fragments by size after electrophoretic separation. A particular difference in sequence between the two or more original DNA molecules is revealed by differences in the length of the fragments, provided that the difference is in the recognition site of the enzyme thereby blocking its activity. This principle has revolutionized genetics when it was combined with Southern hybridization for the analysis of genomes of higher organisms (Southern, 1975; Fig. 1.1).

Southern blot hybridization analysis of restriction enzyme digested genomic DNA has been instrumental in the detection of DNA sequence variations, referred to as restriction enzyme length polymorphisms or RFLPs (Flavell et al., 1978; Kan and Dozy, 1978; Jeffreys, 1979). This source of polymorphism has been shown to be very rich even to an extent where it is possible to construct genetic maps of chromosomes based on RFLPs (Botstein et al., 1980; White et al., 1985). The chances of detecting unknown DNA sequence variation purely on the basis of altered restriction enzyme recognition sites are quite slim. Nevertheless, DNA sequence variations have been detected in this way (e.g. Maddalena et al., 1988; Carothers et al., 1988). Based on such analyses the human genome was estimated to contain 1 in 300 (autosomes) to 1 in 1000 (X chromosome) polymorphic nucleotides (Barker et al., 1984; Hofker et al., 1986). The method was also the first to be used to detect a disease causing point mutation for sickle cell anaemia (Chang and Kan, 1982; Orkin et al., 1982). At present it is still used in combination with other techniques, like PCR, for the detection of particular mutations, such as the sickle cell mutation in human DNA (Saiki et al., 1985) and the bovine leukocyte adhesion deficiency (BLAD) mutation in cattle DNA (Shuster et al., 1992).

Although the mutation scanning capacity of RFLP analysis by itself is low, restriction enzyme digestion is useful for the generation of fragments which can be further separated by electrophoretic methods described below. In addition, it can be used in two-dimensional separations in which a second restriction enzyme digestion is performed after the first size separation and followed by a second size separation.

1.2.3 Analysis of basepairing

The structural basis of the double-stranded DNA molecules as discovered by Watson and Crick (1953a,b) is their sequence-dependent basepairing and stacking characteristics: A can only perfectly pair with T, and C only with G. Deviations from this Watson–Crick type of basepairing can form the basis of the detection of DNA sequence variation within a particular sequence. Such sequence-specific deviations from perfect basepairing can be detected as changes in hybridization characteristics of a probe to a target DNA molecule, changes in melting (strand separation) behaviour of a double-stranded molecule or as vulnerability of heteroduplexed DNA molecules to react with nucleases or particular chemicals.

1.2.3.1 Hybridization analysis

This method for detecting sequence variation is based on whether or not there is formation of a heteroduplex molecule (a double-stranded DNA molecule in which there is incomplete basepairing owing to a mismatch at one or more positions in the sequence) between a labelled nucleic acid probe and a target under certain hybridization conditions. In the usual format of this assay the target sequence is made single stranded by denaturation, after which a single-stranded labelled probe molecule is added. If there are more than a certain number of basepair mismatches within the heteroduplex the probe will not bind and hence no signal will remain after removal of unbound probe. This way of detecting sequence variations using polynucleotides is very insensitive (typically, 70% homology in basepair sequence between probe and target is sufficient for probe hybridization). However, it can for instance be applied for so-called 'zoo blot' hybridization analysis where, in the search for homologous DNA sequences in different species, large deviations from the wild-type DNA sequence can be expected.

The hybridization principle was later employed in refined form by using oligonucleotides to detect the presence or absence of point mutations (Wallace et al., 1979). In this method two oligonucleotides are synthesized, complementary to either the wild type sequence or to a particular mutant sequence, labelled and used as hybridization probes on Southern blots or dried gels of genomic DNA separation patterns (Conner et al., 1983; Piratsu et al., 1983) or PCR-amplified sequences spotted on filter (Saiki et al., 1986 and 1988). Examples of mutations detected through this method include mutations in the β-hexosaminidase A gene (Hex A; Paw et al., 1990) and lipoprotein lipase gene mutants (Emi et al., 1990).

Because of the exquisite sequence specificity of the hybridization reaction only one oligonucleotide can normally be applied at a time. Tetra-alkylammonium salts, however, can reduce the dependency of the melting temperature on the base composition (Melchior and Von Hippel, 1973; Wetmur et al., 1981). In this way identical hybridization conditions can be applied for several different oligonucleotides in the so-called reverse dot blot assay (Saiki et al., 1989). This can be especially advantageous when screening genes for several different mutations. It is expected, however, that this will only be possible for the detection of no more than 30 different point mutations in a single hybridization analysis. As such the method lends itself well to mutation screening purposes, that is, to look for the presence of a given (set of) point mutation(s) at a particular location, rather than for scanning areas of the genome for unknown DNA sequence variations.

One particular application of hybridization analysis involves the use of very short oligonucleotides for mapping of ordered libraries of cloned sequences (Lehrach et al., 1990; Lennon and Lehrach, 1991) and for sequencing (Drmanac et al., 1989). In these applications the redundancy of short sequence motifs in the genome is exploited to an extent where information on fragment order and even on the actual basepair sequence can be reconstructed on the basis of a 'hybridization signature' of a particular set of clones using a (large) collection of motifs. In this respect, such an approach can be extended to comparison of genomes in which slight deviations of a wild-type sequence can be detected as aberrations in the hybridization signatures of the corresponding sets of clones (Lehrach *et al.*, 1990).

1.2.3.2   *Solution melting of heteroduplex molecules*

The solution melting method (Smith *et al.*, 1988) exploits the phenomenon that strand separation of duplex molecules, as caused by increased temperature, is sequence dependent and occurs in so-called melting domains (see 1.2.3.4.1). In this technique the target RNA or DNA fragment is solution hybridized to an RNA probe, and the resulting RNA–RNA or RNA–DNA heteroduplexes are subsequently exposed first to tetraethylammonium bromide, which destabilizes dsDNA and RNA molecules and facilitates helix-coil transitions, and subsequently to a step-wise formamide gradient in different tubes. The target molecules will be partially double-stranded only at the site of the so-called highest (or latest) melting domain so that at particular formamide concentrations the highest melting domain in the duplex molecule will completely dissociate and single-stranded fragments will arise. These resulting fragments are then analysed by polyacrylamide gel electrophoresis to distinguish between single-stranded (slow migrating) and double-stranded (fast migrating) RNA or DNA molecules. Sequence variants of the highest melting domain in the fragment of interest can be discerned by the different formamide concentrations at which they will display complete strand dissociation. Under appropriate conditions even single base substitutions can be detected. The method has the advantage that sequence variants in high melting domains can be detected (of up to 140 bp fragments for RNA–DNA heteroduplexes and up to 260 bp fragments for RNA–RNA duplexes) but considerable ambiguity has been observed in distinguishing homozygotes from heterozygotes, thereby effectively limiting the analysis to X-linked genes (Latham and Smith, 1989). Moreover, the application of GC-clamped PCR (Sheffield *et al.*, 1989) has proven easier to analyse high melting domains (see 1.2.3.4.1).

1.2.3.3   *Basepair mismatch reactivity of heteroduplex molecules*

The aberrant conformation of a nucleic acid heteroduplex molecule at the site of a basepair mismatch can be the target for either enzymatic or chemical reactions. These can form the basis of methods for finding DNA sequence variants through the formation of heteroduplexes. One distinguishes between enzymatic reactions, based on S_1 nuclease and RNase A digestion, and chemical reactions, based on carbodiimide modification and chemical cleavage of mismatched basepairs.

1.2.3.3.1   *S_1 nuclease digestion*

S_1 nuclease recognizes and digests single-stranded DNA and has also been shown to recognize single-stranded DNA regions in heteroduplex RNA–DNA and DNA–DNA molecules (Shenk *et al.*, 1975). Although sensitivity to single base mismatches has been reported (Shenk *et al.*, 1975), in general, only mismatches encompassing at least 4 basepairs can be reliably detected (Dickson *et al.*, 1984; Chebloune *et al.*, 1984).

1.2.3.3.2 RNase A digestion

Similar to S_1 nuclease, ribonuclease A (RNase A) has been shown to recognize single-stranded RNA in RNA:DNA duplexes (Myers *et al.,* 1985e, 1989) and in RNA:RNA duplexes (Winter *et al.,* 1985). The method involves the generation of radio-labelled single-stranded RNA probes and their subsequent hybridization to a single-stranded (denatured) DNA target, followed by digestion with RNase A. The method is sensitive to single base mismatches but some of these, depending on the sequence context of the mismatch, have been shown to be refractory to RNase A cleavage (Myers *et al.,* 1985e). By analysing each of the two strands of a particular DNA fragment with their corresponding RNA probes (of opposite sense) in separate RNAse A reactions, it is possible to detect up to 70% of all possible sequence changes in the DNA fragment. It appears that the presence of purines at the mismatch prevents such mismatches from being cleaved (Cotton, 1993). The method has been successfully used, e.g. for the detection of β-globin mutations in human genomic DNA (Myers *et al.,* 1985e), cKi-ras oncogene variants in RNA from tumour cell lines (Winter *et al.,* 1985), and previously unknown mutations in the hypoxanthine phosphoriboxyl transferase (HPRT) gene in Lesch–Nyhan patients (Gibbs and Caskey, 1987) and an unknown mutation in the ornithine transcarbamoylase (OTC) gene in mouse genomic DNA (Veres *et al.,* 1987).

1.2.3.3.3 Carbodiimide modification

The reactivity of mismatched basepairs with carbodiimide was initially exploited in a mobility shift assay in polyacrylamide gels (Novack *et al.,* 1986). In this method heteroduplexed DNA molecules are allowed to react with carbodiimide, which specifically reacts with unpaired G and T residues and results in the presence of a tag at the site of the mismatch. Sequence variants can then be detected by means of their altered gel electrophoretic mobility caused by the presence of the carbodiimide tag.

More recently, this tagging system of heteroduplexed molecules was combined with the application of antibodies specific for carbodiimide and used to detect mutations in very large fragments (up to 7 kb) by immunomicroscopy (Ganguly *et al.,* 1989). Another variation of the method involves the use of heteroduplexed and carbodiimide-tagged DNA molecules in a subsequent PCR reaction under conditions in which primer extension terminates at the site of the modified bases. This will lead to the production of additional shorter PCR products (Ganguly and Prockop, 1990). This method has been used to detect a new mutation in the human type I procollagen (COL1A1) in an osteogenesis imperfecta patient (Zhuang *et al.,* 1991).

1.2.3.3.4 Chemical cleavage of basepair mismatches (CCM)

The base-specific reaction of particular chemicals with DNA molecules has been first exploited in DNA sequencing (Maxam and Gilbert, 1977; see 1.2.1). This principle, however, can also be used to establish the presence of sequence variations in heteroduplexed molecules which contain basepair mismatches. Two reagents, osmium tetroxide and hydroxylamine, were found to modify mispaired

T and C, respectively. When followed by piperidine-catalysed cleavage, the size of the resulting fragments can give positional information on the mismatch (Cotton, *et al.,* 1988; Cotton and Campbell, 1989). Although particular mismatches have been shown to be cleaved less efficiently, analysis with probes of opposite sense allows nearly 100% of possible mismatches to be detected (Dianzani *et al.,* 1991a; Forrest *et al.,* 1991). The method has been successfully applied in the detection of sequence variations in human genes, e.g. the coagulation factor IX gene (Montandon *et al.,* 1989), the β-globin gene (Dianzani *et al.,* 1991a), the phenylalanine hydroxylase (PAH) gene (Dianzani *et al.,* 1991b), the β-hexosaminidase A gene (Akli *et al.,* 1991), the OTC gene (Grompe *et al.,* 1991) and the dystrophin gene (Roberts *et al.,* 1992).

1.2.3.4 Gel electrophoretic mobility assays

Many of the above-mentioned methods involve the electrophoretic size separation of DNA fragments as a secondary means of identifying DNA sequence variation. Below, electrophoretic separation methods will be discussed based on different separation criteria and exploiting sequence-dependent variations in basepairing which cause differences in conformation and lead to altered electrophoretic mobilities. In these electrophoresis methods the induced local loss of interstrand basepairing leads to the arising of molecules with a conformation which is very different from that of the normal worm-like rods. Because the migration of DNA molecules in polyacrylamide (PAA) gels is dependent on their conformation, this in turn determines the position in the gel where DNA molecules undergo severe mobility retardation.

At increased temperature, differences in melting behaviour of duplex molecules can form the basis of separation of DNA sequence variants by gel electrophoresis. Such separations are based on the formation, before or during electrophoresis, of conformation differences between the sequence variants, which will lead to different electrophoretic migratory behaviour of the variants. In this way the melting (strand-separation) characteristics of double-stranded DNA molecules or DNA–RNA or RNA–RNA hybrid molecules can be assessed. Thermosensitive gel electrophoretic separations are based on the use of temperature gradients applied in the gel, either as a true temperature gradient (temperature gradient gel electrophoresis: TGGE) or as a gradient of denaturants such as urea and formamide (denaturing gradient gel electrophoresis: DGGE).

In techniques other than DGGE or TGGE altered electrophoretic mobility as such is measured; this altered mobility results from differences in intra-strand basepairing (as in SSCA: single-strand conformation analysis; see 1.2.3.4.3), or by the presence of a basepair mismatch (heteroduplex analysis). Gel electrophoretic separation of homoduplex double-stranded DNA molecules under neutral conditions shows very little, if any, sensitivity to the base sequence of the molecules but rather to their length. Under special electrophoretic conditions the analysis of either single-stranded DNA molecules or heteroduplexed DNA molecules can provide the means to detect deviations from a wild-type sequence.

Examples of the direct dependency of electrophoretic mobility on DNA conformation differences are SSCA and the analysis of heteroduplex molecules.

1.2.3.4.1 Denaturing gradient gel electrophoresis (DGGE)

Principles. DGGE involves separation in a polyacrylamide gel containing a gradient of denaturants (i.e. urea and formamide) at a fixed and elevated temperature (Fischer and Lerman, 1979a,b). The double-stranded DNA fragments will, at first, travel through the gel according to their size, until denaturant concentrations are encountered inducing the strands to leave their helical state and to dissociate or melt. The strand dissociation is length independent and highly sequence dependent, owing to differences in basepairing and the influence of neighbouring bases on stacking forces in the double-helical chain (Fischer and Lerman, 1979a, 1980, 1983). Once part of the molecule is melted, the fragment undergoes a change in conformation and a concomitant reduction in electrophoretic mobility (for reviews, see Lerman *et al.,* 1984, 1986).

Melting of DNA fragments is not a gradual continuous process but proceeds in discrete steps, owing to the presence of so-called melting domains: stretches of roughly 50–400 basepairs with nearly identical melting temperatures. Strand separation of the domain with the lowest melting temperature (the lowest- or first-melting domain) initiates the mobility drop. This migratory reduction, however, is critically dependent on the presence of a juxtaposed higher melting domain (referred to as a 'clamp'), the strands of which are still helical at the melting temperature of the first melting domain. At that particular concentration of denaturants a branched molecule will arise with a much lower mobility in the polyacrylamide matrix. Therefore, at a certain position in the denaturing   gradient, migration of a particular fragment in the DGGE gel is no longer size dependent but is dictated by the basepair sequence of the first melting domain.   For calculation of the melting temperature of a particular basepair sequence a computer algorithm is available (Fischer and Lerman, 1979a; Lerman *et al.,* 1984, 1986). Based on thermodynamical stability values of a given basepair with its neighbouring basepairs, the computer program can calculate the $T_m(50)$ value for each basepair in a particular sequence. $T_m(50)$ is the temperature at which the melting domain is completely single stranded in 50% of the fragments. Since particular concentrations of denaturants in the DGGE gel correspond to particular temperatures, $T_m(50)$ positions of particular fragments in the gel can be calculated using parameters such as buffer concentration and run time. The computer program has been shown to be quite accurate in predicting the gel positions of particular fragments and their sequence variants, even when they differ by only a single point mutation including A→T and G→C substitutions (Myers *et al.,* 1985b, 1985c; Lerman *et al.,* 1986; Abrams *et al.,* 1990). In reverse, fragments observed at a certain position in the gel can be characterized by a certain $T_m(50)$. This feature, however, has not been investigated to an extent allowing sequence variations to be deduced from particular positions of a given fragment in a DGGE gel.

In Fig. 1.2 an example of the application of the MELT program and the detection of point mutations by DGGE is shown for four variants of a 192 bp cloned DNA fragment derived from bacteriophage mu (Uitterlinden and Vijg, 1989). The fragments, schematically depicted in Fig. 1.2a, differ only by point mutations at the 3′ end of the molecule. Therefore, only the T_m50 of the lowest melting domain (positions 160 to 192 at the 3′ side of the molecule) differs (Fig. 1.2b). On the basis of these differences the order of melting can now be predicted to occur in the following sequence: first, fragment 7, then 8 and 6, and finally the wild type (649). This is in keeping with the nature of the mutations. That is, they all are G→A substitutions and hence substitute the relatively stable G–C basepair for its more unstable counterpart: the A–T basepair. Indeed, fragment 7 with two of these substitutions can be expected to have the lowest melting temperature.

To study the correspondence between the melting behaviour as predicted by the computer algorithm and the actual electrophoretic behaviour in DGGE gels, the fragments are subjected to perpendicular DGGE. In this form of DGGE the gradient of denaturants concentration is perpendicular to the direction of electrophoresis (Fig. 1.2c). The fragments to be analysed are loaded in a single slot spanning the complete width of the gel. In this way, the exact point in the gradient where melting of the lowest melting domain in each of the fragments occurs can be determined. In Fig.1.2c inflections for each of the four fragments are indicated. The sequence of melting is exactly as predicted by the algorithm. Note that fragments 6 and 8, which initially co-migrate as one line, split up at the end of the second inflection.

To exploit this melting behaviour for scanning purposes parallel DGGE analysis is applied (Fig. 1.2d). The four fragments migrate to identical positions during electrophoresis in a neutral polyacrylamide gel (Fig. 1.2d, left), but to different positions in a denaturing gradient gel (Fig. 1.2d, right). The migration in DGGE corresponds to the order predicted by the algorithm (Fig.1.2b).

Thus, the sequence-dependent separation in DGGE gels is based mainly on three characteristics of double-stranded DNA molecules: (A) melting (or strand separation) of contiguous bases in DNA molecules is closely coupled to form so-called melting domains; (B) the temperature (or the concentration of denaturants, such as urea and formamide) at which strands of a melting domain part is highly dependent on the sequence composition of the domain; (C) partially melted DNA molecules (containing both double-stranded and single-stranded regions) have a decreased electrophoretic mobility in PAA gels in comparison with native, double-stranded molecules.

Analysis of genomic DNA. An important extension of the DGGE system is its application to the analysis of total genomic DNA through the use of PCR amplification (Cariello *et al.*, 1988a; Sheffield *et al.*, 1989) or by transfer of separation patterns to hybridization membranes. The latter can be accomplished by capillary blotting using modified gel media (Børresen *et al.*, 1988) or by means of electroblotting. The latter was first described for the analysis of micro- and minisatellite repeat motifs using core probes (Uitterlinden *et al.*, 1989a) and

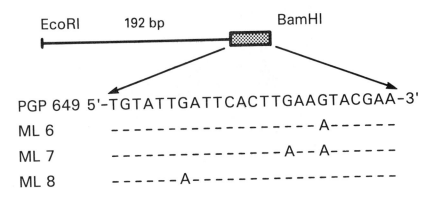

PGP 649 5'–TGTATTGATTCACTTGAAGTACGAA–3'

ML 6 — — — — — — — — — — — — — — — — — — — A — — — — — —

ML 7 — — — — — — — — — — — — — — — A — — A — — — — — —

ML 8 — — — — — — A — — — — — — — — — — — — — — — — — —

(a)

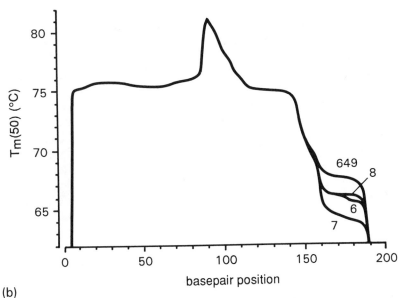

(b)

Fig. 1.2. (a) Schematic depiction of four different sequence variants of a 192 bp Eco RI-Bam
H1 fragment derived from bacteriophage mu. (b) Composite of the melting maps of the four
variants shown in (a). Note that the sequence variation is present in the lowest melting domain
at the 3′ end (from Uitterlinden and Vijg, 1989).

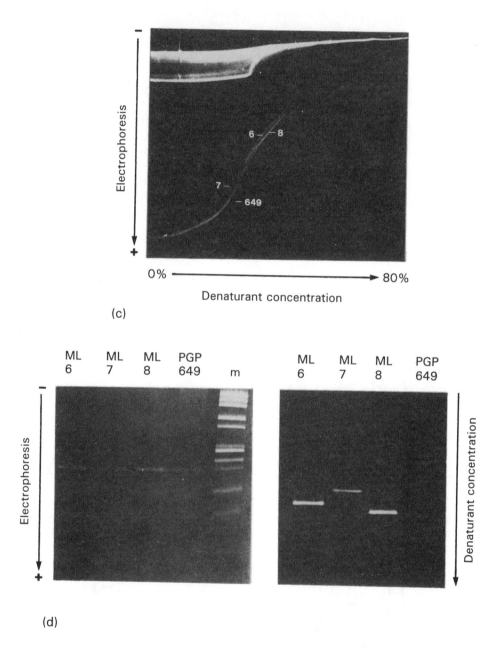

(c)

(d)

Fig. 1.2.(c) Perpendicular DGGE analysis of a mixture of the four sequence variants shown in (a). The intense band in the top of the gel is the plasmid vector DNA from which the fragments were excised. The particular sequence variants, indicated by their numbers, were loaded in unequal concentrations to identify them. (d) Neutral polyacrylamide gel electrophoresis (left) and parallel denaturing gradient gel electrophoresis (right) of the four sequence variants shown in (a). m = size marker locus-specific VNTR probes (Uitterlinden and Vijg, 1991). More recently, electroblotting has been extended to the detection of single-copy DNA sequences in *Drosophila* (Gray *et al.*, 1991) and in humans (Burmeister *et al.*, 1991; Gray 1992).

GC-clamping. The attachment of a GC-rich sequence (referred to as a 'clamp') to a fragment which normally undergoes immediate strand dissociation on melting has been shown to alter the melting characteristics of the fragment (Lerman *et al.*, 1984; Myers *et al.*, 1985b). The clamp prevents monodomain fragments from completely dissociating and guarantees the adjacent sequence always to be the first melting domain. When a fragment is 'clamped', partially melted molecules can arise on melting and hence allow the comprehensive detection of mutations in the fragment of interest (Myers *et al.*, 1985c). This principle has been applied in the improvement of detection of sequence variations in genomic DNA through PCR with 'clamped' primers (Sheffield *et al.*, 1989, 1992a and 1992b), through selective attachment of a clamp to genomic restriction fragments (Abrams *et al.*, 1990) or by 'chemical clamping' involving the chemical modification of the outer basepair of a PCR-primer to link the opposite bases covalently (Costes *et al.*, 1993).

Analysis of heteroduplex molecules. Although DGGE is capable of detecting single base differences by comparing the naturally occurring homoduplexes, its sensitivity can be even further increased by introducing base pair mismatches as they occur in heteroduplexes (Lerman *et al.*, 1984). This can be illustrated by the analysis of a point mutation in the first exon of the cHa-ras1 proto-oncogene (Uitterlinden and Vijg, 1990). The mutation is a G→T substitution which was found in T24 bladder carcinoma cells (Premkumar Reddy *et al.*, 1982). In Fig. 1.3a the schematic organization is shown of the entire cHa-ras1 proto-oncogene as well as its melting map. Several features can be discerned in the $T_m(50)$ plot, such as the HTF island representing a very high melting domain and the polyadenylation signal region which stands out as a small dip in $T_m(50)$. Strikingly, large parts of all four exons are low melting domains.

Separate melting maps calculated for the first exon of the wild type and the T24 sequence variant are shown in Fig. 1.3b. These fragments correspond to positions 1640 to 1853 (213 bp) of the gene sequence of the GENBANK entry and include a 21 bp piece of the multiple cloning site of the pTZ18R vector in which these parts were cloned. The 234 bp Pst I inserts contain the normal and mutated T24 sequence of the first exon, referred to as pN and pT, respectively. They were subcloned from the corresponding HOMER cosmid clones (Spandidos and Wilkie, 1984). The lowest melting domain spans base positions 20 to 150 for both plasmids but the $T_m(50)$ value of this domain is 0.4°C lower for the pT Pst I insert owing to the presence of the T at position 79 in this domain. Fig. 1.3c shows the perpendicular DGGE analysis of the Pst I insert of pN. A major inflection at 48% denaturant concentration corresponds to melting of the lowest melting domain. The fragment apparently remains partially single stranded, indicating the possibility of separating sequence variants by parallel DGGE.

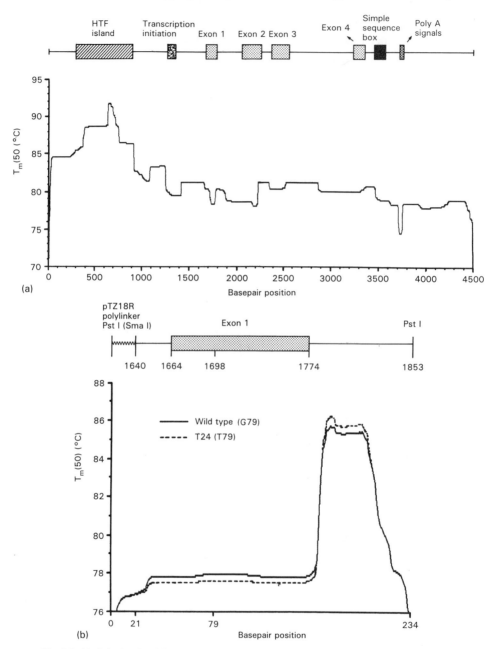

(a)

(b)

Fig. 1.3. (a) Calculated melting map of the human cHa-ras1 proto-oncogene as published by Capon *et al.* (1983) using basepairs 0 to 4500 from the GENBANK entry (HUMHRAS1). HTF=Hpa II tiny fragments. (b) Calculated melting map of the first exon of the human cHa-ras1 proto-oncogene and the T24 variant sequence as present in the 234 bp Pst I fragments used for DGGE analysis. The T at position 1698 in the T24 oncogene corresponds to position 79 in the Pst I insert and lowers the melting temperature of the first or lowest melting domain (from Uitterlinden and Vijg, 1990).

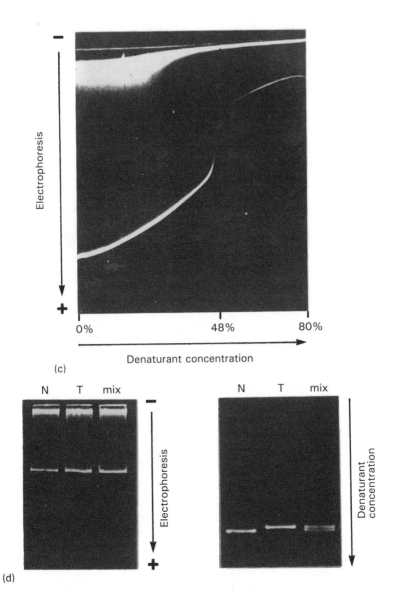

(c)

(d)

Fig. 1.3. (c) Perpendicular DGGE analysis of the cloned 234 bp Pst I fragment containing the first exon of the human cHa-ras1 proto-oncogene. The intense band in the top of the gel is the plasmid vector DNA from which the Pst I insert was excised. A major inflection at 48% can be observed. Electrophoresis was performed at 60°C at 150 V for 2.5 h in a 1 × TAE 6% PAA gel. (d) Parallel electrophoresis of the 234 bp Pst I fragment containing the first exon of the human cHa-ras1 proto-oncogene. N stands for the wild-type sequence and T for the T24 variant. On the left results are shown for neutral 6% polyacrylamide gel electrophoresis, run at 150 V for 1 h at 60°C, and on the right the DGGE separation pattern. DGGE was performed at 60°C at 150 V for 8 h in a 1 × TAE 6% PAA gel containing a 35–65% UF gradient. Mix is a mixture of the wild type and the T24-derived 234 bp Pst I fragment.

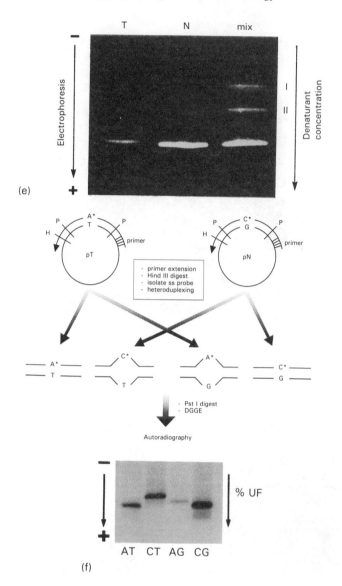

Fig. 1.3. (e) Parallel DGGE analysis of the heteroduplex molecules of the 234 bp Pst I fragments. N stands for the wild-type homoduplex sequence, T for the T24 homoduplex sequence and mix stands for the heteroduplex mixture. Note that in this experiment the two homoduplex molecules are not separated, probably as a result of overloading. I indicates the CT-mismatch-containing heteroduplex and II the AG-mismatch-containing heteroduplex. DGGE was as described in (d). (f) Top: schematic representation of the procedure followed to obtain single-stranded, radiolabelled probe derived from the wild type and T24 mutant 234-bp Pst I fragment. H=Hind III recognition site; P=Pst I recognition site. (see Uitterlinden and Vijg (1990) for details). Bottom: autoradiograph of the DGGE analysis of the homo- and heteroduplex molecules formed using the labelled single-strand probes for the 234 bp Pst I fragment. Lettering below the lanes refers to the basepair at position 79.

We subsequently analysed pN and pT Pst I fragments on a 35–65% parallel DGGE gel. As shown in Fig. 1.3d the normal and T24-derived fragments migrate to slightly different positions in the gradient after electrophoresis. The pN migrated 2 mm further in the gel than the pT Pst I fragment, which was sufficient to separate both fragments in a mixture (right lane in Fig. 1.3d). Since the DGGE gel system is exquisitely sensitive for mismatches in heteroduplexed molecules (Lerman *et al.*, 1984) we performed heteroduplex analysis of the two Pst I inserts. Fig. 1.3e shows the parallel DGGE gel of the heteroduplexed mixture. In addition to the homoduplex bands we observed two extra bands which were halted earlier in the gel as a result of the presence of mismatches. In order to identify these bands we performed heteroduplex analysis with labelled single strands derived from both pN and pT plasmids. Fig. 1.3f is a schematic representation of the production and sequence composition of the labelled strands and the annealing products. In Fig. 1.3g the DGGE analysis of the heteroduplex experiment is shown. The CT mismatch is less stable than the AG mismatch, as demonstrated by the position of the CT mismatch band at lower denaturants concentration (i.e. higher in the gel) than the GA mismatch. From this we infer the upper band in the plasmid heteroduplex mixture (I in Fig. 1.3e) to be composed of CT mismatch containing molecules and the lower band (II in Fig. 1.3e) to contain the AG mismatch containing molecules.

Heteroduplexing in DGGE has also been exploited in the analysis of genomic DNA to increase the percentage of mutations detected. This includes heteroduplexes formed between genomic DNA restriction fragments and radiolabelled single-stranded DNA probes for human DNA (Myers *et al.*, 1985a; Noll and Collins, 1987; Bewsey *et al.*, 1991) and for maize plant DNA (Riedel *et al.*, 1990). The same principle has been demonstrated to improve mutation detection efficiency for RNA : DNA heteroduplexes in human DNA (Takahashi *et al.*, 1990) and for DNA : RNA and RNA : RNA heteroduplexes in the analysis of mutation rates of viral genomes (Smith *et al.*, 1986; Leider *et al.*, 1988).

CDGE. Increased resolution of sequence variants of particular fragments can be obtained by applying so-called constant denaturing gel electrophoresis (CDGE; Hovig *et al.*, 1991). Here a constant denaturant concentration is applied in the gel which corresponds to a specific melting domain within the fragment of interest. Particular advantages are that increased separation distances are obtained and that the system is more amenable to automation than DGGE. However, disadvantages are the time dependency of the separation and the limitation to only one single fragment. A high sensitivity of CDGE in detecting point mutations was shown for the HPRT gene in mice and hamster DNA (Hovig *et al.*, 1991) and the p53 gene in human tumour DNA (Børresen *et al.*, 1991).

Analysis of complex populations of molecules. A particular advantage of DGGE in comparison with other mutation detection techniques is its potential to identify simultaneously different sequence variants in a complex population of molecules.

Such populations can consist of a wild-type molecule and mutants present at different abundances or of a mixture of closely related molecules derived from different organisms. One important application of DGGE in characterizing the former population has included the detection of sequence variants after treatment of DNA with damaging agents (Myers *et al.*, 1985d; Thilly, 1985; Lillie *et al.*, 1986; Cariello *et al.*, 1988b, 1991a). In most of these schemes DNA fragments, prepared from cells which had been in contact with a mutagen, are amplified by means of PCR and subsequently separated on a DGGE gel, optimized for separating variants of the fragment of choice. This application makes use of unique advantages offered by DGGE (or TGGE for that matter; see below). These include the possibility of isolating the mutant sequences in an intact form and detecting low frequency mutation events in populations of molecules. The results obtained so far indicate that the fidelity of the amplification step is limiting the sensitivity, owing to errors which are introduced by the DNA polymerase (Cariello *et al.*, 1991b). Enzymes with higher fidelity than the original Taq polymerase have recently been described, such as Sequenase, Vent, and *Pfu* polymerase. It appears, however, that even with these high fidelity enzymes a mutation must be present in about 1% of the cells in the original cell population to be detectable (Cariello *et al.*, 1991b). In a similar way Keohavong *et al.* (1991) used DGGE to detect point mutations in exon 3 of the HPRT gene in mass cultures of mutant cells (i.e. 6-thioguanine resistant) which were exposed to UV light.

A similar application of DGGE analysis in this respect is the analysis of complex populations of individuals, such as those found in bacterial mats. These mats represent ecological niches in which a variety of different organisms are present in different numbers. Such populations can be typed by means of DGGE analysis of their 16S small subunit ribosomal RNA (16S ssu-rRNA) genes (Muyzer *et al.*, 1993). For this purpose, a particular region from the 16S RNA (the V4 region) in which species-specific variation is expected is PCR amplified from genomic DNA isolated from a particular mat and subjected to DGGE analysis. The amplification of the 233 bp 16S fragment includes the attachment of a 40 bp GC-clamp to ensure optimal melting behaviour in DGGE. Fig. 1.4a shows a so-called time-travel experiment in which the PCR amplified 16S fragments derived from two different bacterial species, *Desulfovibrio desulfuricans* and *Escherichia coli*, were loaded every 10 min on the DGGE gel. The fragments show a dramatic decrease in mobility at a species-specific position in the DGGE gel. This position corresponds to melting of the lowest melting domain of the fragment. Because of sequence differences among these domains the fragments can be effectively separated by DGGE.

When 16S fragments derived from five different bacterial species are analysed in this way, good separation can be obtained (Fig. 1.4b). This is also true when their genomic DNAs are mixed before the PCR amplification, a situation similar to real microbial populations. Fig. 1.4c shows that a band derived from a particular species can still be detected, even when the species constitutes only 1% of the original population.

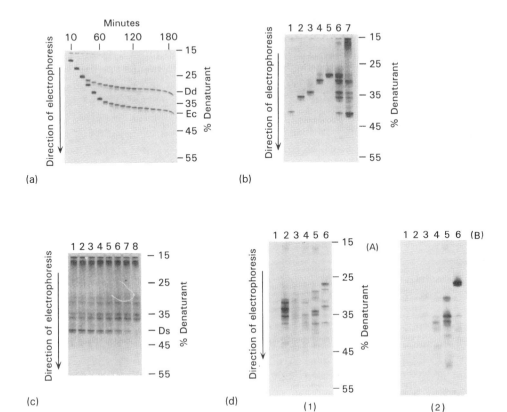

Fig. 1.4. (a) Negative image of an ethidium bromide stained parallel DGGE separation pattern of a mixture of PCR-amplified 16S DNA fragments from *D. desulfuricans* (Dd) and *E. coli* (Ec) which were loaded every 10 min during a 3 h period. DGGE was performed at 60°C at 200 V in a 6% PAA gel containing a 15–55% UF denaturing gradient. (b) DGGE analysis of PCR-amplified 16S DNA fragments from different eubacteria. The species are *D. sapovorans* (lane 1), *E. coli* (lane 2), *M. chtonoplastes* (lane 3), *T. Thioparus* (lane 4), *D. desulfuricans* (lane 5), a mixture of these PCR products (lane 6), and a sample obtained after PCR amplification of a mixture of the bacterial genomic DNAs (lane 7). The pattern shown is a negative image of an ethidium bromide stained separation pattern. (c) Negative image of an ethidium bromide stained parallel DGGE separation pattern of eight PCR-amplified 16S DNA fragments in which the target DNA from *D. sapovorans* (Ds) was two-fold serially diluted in the PCR solution while the amounts of target DNAs of *E. coli, M. chtonoplastes, T. thioparus,* and *D. desulfuricans* were kept constant. (d) DGGE analysis of 16S DNA fragments obtained after PCR amplification of genomic DNA from uncharacterized microbial populations and individual bacteria. (1) A negative image of ethidium bromide stained parallel DGGE separation patterns of (a) microbial mat samples 1, 2 and 3 (lanes 1, 2, and 3, respectively) and (b) a bacterial biofilm grown under anaerobic conditions (lane 5). Lane 6 contains the separation pattern of a mixture of PCR fragments of five individual bacteria (*D. sapovorans, E. coli, M. chtonoplastes, T. thioparus,* and *D. desulfuricans*). This mixture served as a positive control for the hybridization experiment. (2) Hybridization analysis of the DGGE separation pattern shown in (1) with an oligonucleotide probe specific for sulphate-reducing bacteria (from Muyzer *et al.*, 1993).

The analysis of real microbial populations using this technique shows that indeed 'DGGE fingerprints' can be obtained. As shown in Fig. 1.4d some populations display at least 10 well-resolved bands in the DGGE fingerprint, most likely corresponding to as many different species in the population. The different intensities of the bands observed in the DGGE profile probably correspond to the relative abundances of the particular species in the population. By hybridization analysis of such separation patterns particular groups of bacteria can be highlighted. As illustrated also in Fig. 1.4d the DGGE gel can be transferred to a nylon membrane and hybridized with a probe specific for sulphate-reducing bacteria. This can identify particular bands to be derived from particular species and also enhances the sensitivity of detection.

The procedure described here has general applications. Depending on the primers used particular populations of molecules can be analysed. These can include different bacterial populations, but also populations of higher eukaryotes such as yeast and fungi. The latter application can have important clinical relevance in studying patients suffering from combined infections.

Gene analysis. Since its introduction in 1979, DGGE has been used extensively and, as a result of its exquisite sensitivity for differences in basepair composition (Fischer and Lerman, 1983), it has been widely used for the detection of sequence variations as small as point mutations or even base modifications such as methylation (Collins and Myers, 1987).

In Fig. 1.5 the different steps involved in DGGE analysis of a genomic DNA fragment by PCR are illustrated with examples from the analysis of exon 10 of the cystic fibrosis transmembrane conductance regulator (CFTR) gene. In this case the experimental design was based on PCR amplification of exon 10 in individuals homo- and heterozygous for a neutral sequence variant (G→A point mutation at position 470). The two variants are termed M470 (with A at position 470) and V470 (with G at position 470). First, the melting behaviour of the fragments (derived from a heterozygous individual with both M470 and V470) is analysed by perpendicular DGGE (Fig. 1.5a, left). In total four different fragments are present. First, the homoduplex fragments M470 and V470 result in two, at this broad range of denaturants hardly discernible, lines. The two heteroduplex fragments (G–T and C–A) are byproducts of the PCR and visible as a thin line at slightly lower denaturants concentrations (see also below). In this way, the exact point in the gradient where melting of the lowest melting domain occurs (inflection at 34% UF) and stability of the remaining partially melted fragment (from 34% to 54% UF) are determined. Then, the optimal electrophoretic conditions (time, gradient) are determined in a so-called time travel experiment (Fig. 1.5a, right). Of the two bands observed, the upper one is the fragment of interest while the lower one represents the primers. The fragment can be observed to show strongly reduced electrophoretic mobility at the point in the gradient that corresponds to the concentration of denaturants at the inflection observed in the perpendicular DGGE gel.

CFTR exon 10

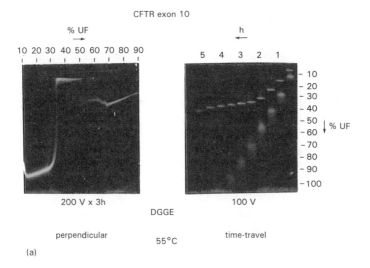

% UF
→

10 20 30 40 50 60 70 80 90

h
←

5 4 3 2 1

- 10
- 20
- 30
- 40
- 50 ↓ % UF
- 60
- 70
- 80
- 90
- 100

200 V x 3h   100 V

DGGE

perpendicular   55°C   time-travel

(a)

CFTR exon 10

homo-   heteroduplex

1 2 3   1 2+3 2 3

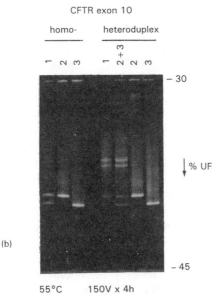

– 30

↓ % UF

(b)

– 45

55°C   150V x 4h

Fig. 1.5. DGGE analysis of a 336 bp fragment encompassing exon 10 of the CFTR gene. (a) Perpendicular and time-travel DGGE analysis of the PCR-amplified CFTR exon 10 fragment obtained from an individual heterozygous for a point mutation at position 470. (b) Heteroduplex analysis. 'Homoduplex' indicates that samples were separated by DGGE immediately after PCR analysis while 'heteroduplex' indicates that the DNAs were first boiled for 10 min and annealed at 60°C for 1 h before loading on the DGGE gel. Individual 1 is heterozygous for a point mutation at position 470 in exon 10, while individuals 2 and 3 are homozygous for the variants M470 and V470, respectively. When genomic DNAs from individuals 2 and 3 are entered in the same PCR reaction a compound DGGE pattern identical to that of the heterozygous individual 1 is obtained, demonstrating the nature of the heteroduplex bands observed.

Fig. 1.5b shows the results of parallel DGGE analysis of PCR-amplified fragments derived from one heterozygous (1) and two homozygous (2 and 3) individuals. The four lanes on the right are from a real heteroduplex experiment. That is, the PCR amplified fragments obtained are denatured and renatured before being subjected to DGGE. Because heteroduplex fragments are inherently more unstable than homoduplex fragments (see Fig. 1.3), this results in the emergence of two early-melting fragments. During PCR amplification heteroduplex molecules are also formed as indicated by the less intense bands in lane 1 on the left (also visible in the perpendicular pattern). Subsequently, the information gathered in this way is utilized to achieve optimal separation of all possible sequence variants (see Chapter 5, section 5.4).

Genes, allelic variants of which have been detected by DGGE include the adenine phosphoribosyl transferase (APRT) gene in the mouse (Dlouhy *et al.*, 1989), and in human DNA the hypoxanthine guanine phosphoribosyl transferase (HPRT) gene (Cariello *et al.*, 1988a), the coagulation factor IX gene (Attree *et al.*, 1989), the cHa-ras1 proto-oncogene (Uitterlinden and Vijg, 1990), the ornithine transcarbamoylase (OTC) gene (Finkelstein *et al.*, 1990), the factor VIII gene (Higuchi *et al.*, 1991), the CFTR gene (Vidaud *et al.*, 1990), the rhodopsin gene (Sheffield *et al.*, 1991), the tumour suppressor gene MCC, p53, prealbumin, rhodopsin, S-antigen, and the transforming growth factor (TGF)-α gene TGF-α (Sheffield *et al.*, 1992b), the insulin receptor gene (Krolewski *et al.*, 1992), the low density lipoprotein (LDL) receptor gene (Top *et al.*, 1992) and the β-globin gene (Losekoot *et al.*, 1990; Rosatelli *et al.*, 1992).

1.2.3.4.2   *Temperature gradient gel electrophoresis (TGGE)*

In modifications of the above-described electrophoretic system, true temperature gradients in combination with a constant urea and/or formamide concentration have been applied instead of using chemical mimics in PAA gels at a constant temperature (Rosenbaum and Riesner, 1987). The method was first applied to the detection of sequence changes in short, double-stranded satellite RNA molecules from viruses (Po *et al.*, 1987; Steger *et al.*, 1987). More recently, however, the system was shown to be capable of detecting single basepair substitutions in DNA molecules (Wartell *et al.*, 1990). Although, in general, the results obtained with TGGE are similar to those found with DGGE, there are differences with respect to the migratory behaviour of DNA molecules. More specifically, an increase of mobility of fragments is observed at temperatures slightly higher than the starting temperature, a phenomenon which is not seen in chemical solvent gradient gels (Wartell *et al.*, 1990).

1.2.3.4.3   *Single-strand conformation analysis (SSCA)*

Under non-denaturing conditions, single-stranded DNA has a folded conformation which is stabilized by intrastrand interactions. The sequence will therefore determine the conformation and hence the electrophoretic mobility. Variation in this mobility of fragments obtained from different individuals is referred to as single-strand conformation polymorphism (SSCP). This principle has been exploited in the electrophoretic separation of fully denatured genomic DNA on neutral polyacrylamide gels followed by transfer of the separation pattern to a nylon membrane and hybridization analysis with RNA probes for each of the strands of a target sequence (Orita *et al.*, 1989a). It has been shown that sequence alterations as small as a single base substitution can be detected in genomic DNA after blot hybridization (Orita *et al.*, 1989a) and after radioactive PCR amplification of the target sequences (Orita *et al.*, 1989b). Electrophoretic conditions (temperature and addition of glycerol) have been shown to be very critical in this method and, according to some reports, only 35% of all point mutations can be detected (Sarkar *et al.*, 1992a; Spinardi *et al.*, 1991).

Recently, Sheffield et al. (1993) reported that the efficiency of mutation detection varies dramatically with the size of the fragment analysed. The optimal size is approximately 150 basepairs in which case 95% of the mutations were detected, but this decreases to less than 10% for fragments of 600 basepairs. Variation in mutation detection efficiency is also influenced by the sequence content of the fragments in that purine-rich fragments allow more different mutations to be detected than purine-poor fragments (Sheffield *et al.*, 1993). Similar to the situation with DGGE, SSCA would gain advantage from a theoretical description   of the influence of intrastrand basepairing on electrophoretic mobility of double-stranded DNA molecules.

SSCA has been successfully applied to detect sequence variants of the ras proto-oncogene and individual Alu repeats (Orita *et al.*, 1989b; Suzuki *et al.*, 1990), the HexA gene (Ainsworth *et al.*, 1991), the PAH gene (Labrune *et al.*, 1991), the CFTR cystic fibrosis gene (Dean *et al.*, 1990; Iannuzi *et al.*, 1991; White *et al.*, 1991; Ivaschenko *et al.*, 1991), the neurofibromatosis gene (Cawthon *et al.*, 1990), the lipoprotein lipase gene (Hata *et al.*, 1990), the factor IX gene (Demers *et al.*, 1990), the KIT proto-oncogene and insulin-like growth factor 1 receptor (IGF1R) gene (Poduslo *et al.*, 1991), the rhodopsin gene (Dryja *et al.*, 1991), and the p53 gene (Runnebaum *et al.*, 1991).

Modifications of SSCA. In a modification of SSCP analysis RNA conformation polymorphisms have been detected by neutral gel electrophoresis of RNA molecules generated from PCR-amplified genomic DNA fragments (Danenberg *et al.*, 1992; Sarkar *et al.*, 1992a). When applied to the scanning of fragments derived from 2.6 kb of the human factor IX gene, RNA-SSCP was shown to detect a higher fraction of point mutations in comparison with DNA-SSCP, i.e. 70% vs 35%, respectively (Sarkar *et al.*, 1992a).

The SSCA mutation detection system has also been combined with dideoxy direct sequencing to provide a screening system for mutations: dideoxy finger-printing or ddF (Sarkar *et al.*, 1992b). In ddF, a ladder of bands of a sequence of interest is generated by performing dideoxy sequencing reactions using one of the four dideoxynucleotides and resolving these on a neutral polyacrylamide gel. Mutation detection can be done along two lines: (a) a position where the dideoxynucleotide can be incorporated is lost or created resulting in loss or appearance of a band in the ladder and (b) one but usually more bands which contain the site of the mutation show different migratory behaviour as in SSCP. When mutations in the factor IX gene were analysed 84 out of 84 different mutations could be detected by ddF (Sarkar *et al.,* 1992b). As the authors stated, however, sensitivity for detecting heterozygosity remains to be established. Furthermore, it was difficult to detect mutations in GC-rich regions owing to compression of bands in the sequencing ladder.

1.2.3.4.4   Analysis of heteroduplex molecules

Direct dependency of electrophoretic mobility on DNA conformation is also exploited in the analysis of heteroduplexed DNA molecules in neutral gel matrices. Aberrant electrophoretic mobilities of heteroduplex molecules have been observed in agarose gels (Shore and Myerowitz, 1990), in PAA gels (Delwart *et al.,* 1993) and in hydrolink gels (Keen *et al.,* 1991). Although few DNA sequences have been analysed by this approach, sensitivity to single basepair substitutions has been reported (Keen *et al.,* 1991; White *et al.,* 1992). Initially, the separation of sequence variants from wild type was rather poor and essentially limited to detection of deletions and/or insertions of several basepairs (see, for example, Paw *et al.,* 1990). More recently, studies have shown that more sequence variants can be detected if gel electrophoretic conditions are optimized by using different gel media and running conditions. These include sequence variants of the HLA locus (Clay *et al.,* 1991), in which most of the variants differed by more than a single basepair, and variants of the rhodopsin gene and the cystic fibrosis gene (Keen *et al.,* 1991). In a study of genetic divergence of HIV-1 *env* genes between HIV strains Delwart *et al.* (1993) observed a reduced mobility in PAA gels of heteroduplexes formed between related sequences proportional to their degree of divergence.

1.2.4   Evaluation

From the previous sections it can be derived that there is at present a wealth of techniques for detecting DNA sequence variation. Among these are mutation scanning methods capable of searching small stretches of DNA and high-lighting any given type of DNA sequence variation. Apart from practical differences, such as varying requirements with respect to skill, laboratory equipment and reproducibility, the different methods have different scanning efficiencies in terms of (a) the number of basepairs that can be scanned per assay and (b) the type of DNA sequence variations that can be detected (Table 1.2). It should be noted that the numbers given for the scanning efficiencies are approximations, since not every method has been tested to the same extent on the same genes.

   DNA sequencing has the highest efficiency in that it can identify any given sequence variant but it is a typical serial approach for identifying DNA sequence variation. RFLP screening is the least efficient mutation scanning method since very few basepair alterations will fall within the palindromic recognition site of a restriction enzyme. Even when large batteries of restriction enzymes are applied very little sequence is effectively scanned for deviations. Restriction enzyme digestion of DNA fragments can be applied in 2-D separation formats in a consecutive way by using a different enzyme before each electrophoretic separation (see section 1.4 on two-dimensional formats).

An advantage of the methods other than sequencing is that they will only identify deviations from the wild-type sequence. The RNAse A and chemical cleavage methods even have the possibility of localizing the site of the variation in the molecule under study. This is accomplished by digestion and/or cleavage at the site of variation and by subsequently measuring the size of the two resulting fragments. However, by doing so these methods do not leave the variant molecules intact for further analysis. The latter is of importance in, for instance, the screening of complex populations of molecules for sequence variants which occur in low frequencies. Furthermore, although no previous knowledge on the sequence of interest is required for the application of RNase A and the chemical cleavage methods, they usually require radioactive labelling and they have been shown to be labour intensive, because several manipulations are required for each sample after PCR amplification, and to lack reproducibility in that not all of the variant molecules are digested/cleaved (Theophilus *et al.,* 1989). However, recent modifications seem to have overcome this problem (Forrest *et al.,* 1991; Cotton, 1993).

Methods in which sequence variants display anomalous electrophoretic mobility have the advantage that the molecules remain intact and can be isolated and purified for further analysis. In this respect, the SSCP method has gained growing popularity despite the fact that it requires radio-active labelling and that only small fragments can be efficiently scanned (up to 90% mutation detection efficiency in 150 bp fragments (Sheffield *et al.,* 1993)). This popularity is due to its ease of use. The heteroduplex mobility assay remains to be evaluated in comparative studies but is unlikely to be able to detect more than 50% of possible sequence variations.

DGGE and TGGE have a high potential for detecting sequence variants and are non-radioactive methods. Apart from only displaying deviations from the wild type and leaving the variant molecules intact, they have the additional advantage that, after separation, the variant and wild-type molecules are widely spaced apart (centimetres rather than millimetres; see Figs 1.2d, 1.4a and 1.5b) in comparison with the other electrophoretic mobility assays. This is of particular importance when this method is incorporated as a separation criterion in a 2D format since it allows most of the 2D gel space to be used for displaying fragments (see section 1.4 on 2D formats). In this respect, DGGE (and TGGE) have the additional advantage that fragments are retarded and hence are much less dependent on the duration time of electrophoresis, in contrast to CDGE in which the separation is critically time-dependent. Finally, in practical terms, DGGE is cheaper and more flexible than TGGE because of the more expensive instrumental requirements for TGGE. In addition, with

TGGE only one gel can be run while in DGGE multiple gels can be handled and run simultaneously.

1.3   MEASURING DNA SEQUENCE VARIATION ON A TOTAL GENOME SCALE

As has been mentioned at the beginning of this chapter, a major aspect of the science of genetics is the search for genetic variation. This has now mainly become a search for variation in DNA. The methods reviewed above allow individual variation in the DNA sequences of the genome to be studied. However, while at one end of the spectrum of methods the sequencing of more than one individual genome cannot be seriously entertained, the methods for measuring sequence variation in multiple individuals simultaneously are limited to very small regions of the genome. The latter alternative for screening populations of individuals for DNA sequence variants is nevertheless the only one feasible.

To assess individual DNA sequence variation on a total genome scale it is necessary first to generate fragments which are smaller than the original DNA molecules, i.e. the chromosomes. Two enzyme-based methods exist nowadays to generate DNA fragments from a genome under study: (1) restriction enzyme digestion, and (2) PCR amplification. Fragments generated in this way can be entered into the mutation scanning analysis.

For the subsequent screening of individuals for sequence variation two different strategies can be followed: a serial one and a parallel one. In a serial strategy sequence information that is available will be used to generate probes and/or primers for mutation scanning of their target sequences. The actual analyses are carried out sequentially, for example, by robotics. The advantage of this strategy is its simplicity; each fragment for which sequence information is available can be analysed. A disadvantage is that in spite of robotics the large numbers of separate determinations are still a formidable problem. For example, for DNA sequencing it can be calculated that, assuming 300 bp to be determined in a single sequencing reaction, at least 50 million sequencing reactions have to be performed to obtain the sequence of only a single human genome (including correction for overlap but without correction for redundancy). In principle, all of the methods mentioned above allow the scanning of individual genomes for DNA sequence differences. However, their application for this purpose is severely hampered by the fact that they have to be applied serially in large-scale studies, such as comprehensive mutation analysis of entire genomes or genomic regions. Indeed, even the scanning of single genes can already be problematic in view of the length of the region to be scanned and the number of different mutations which can occur (see Tables 1.3 and 1.4). An example of the comprehensive analysis of a large gene for mutations is the scanning of the factor VIII gene region in Haemophilia A patients with unknown mutations (Higuchi et al., 1991). In this study 45 different primer sets were necessary to cover 99% of the coding part of the gene (6.9 kb), 41 out of 50 splice junctions and some regulatory

sequences. By applying DGGE analysis these authors were capable of detecting
25 out of 29 previously unknown point mutations in this 187 kb large gene.

Table 1.3. Size of genes in the human genome[a]

Gene	Genomic size (kb)	mRNA (kb)	Number of introns
Small			
α-Globin	0.8	0.5	2
β-Globin	1.5	0.6	2
Insulin	1.7	0.4	2
Apolipoprotein E	3.6	1.2	3
Parathyroid hormone	4.2	1.0	2
Medium			
Protein C	11.0	1.4	7
Collagen I pro-α-1	18.0	5.0	50
Collagen I pro-α-2	38.0	5.0	50
Albumin	25.0	2.1	14
HMG CoA reductase	25.0	4.2	19
Adenosine deaminase	32.0	1.5	11
Factor IX	34.0	1.4	7
Catalase	34.0	1.6	12
LDL receptor	45.0	5.5	17
Large			
Phenylalanine Hydroxylase	90.0	2.4	12
Very large			
Factor VIII	187.0	7.0	26
Thyroglobulin	>300	8.7	>36
CFTR (cystic fibrosis)	250	6.5	27
Extremely large			
Dystrophin (DMD)	>2400	14.0	78

[a] Adapted from McKusick, (1990).

   Studies such as gene identification by linkage analysis, genome scanning of
tumours and other forms of comparative analysis of closely related genomes are
even more demanding and represent a considerable number of handlings. This
problem can be solved by robotics; in practice this turned out to be hardly faster.
This type of serial analysis is therefore, by its nature, rather money, time and
labour consuming and allows information only to be compiled after many separate analyses.

Table 1.4. Mutations in human disease genes

Disease[a]	Fragile X	AAT	CF	DMD	Haem A	Haem B
Incidence patients	1:1250($\male$)	1:1600	1:3600	1:3500($\male$)	1:10 000($\male$)	1:30 000($\male$)
Incidence carriers	1:20	1:25	—	—	—	—
Chromosome	Xq27.3	14q24.3	7q31	Xp21	Xq28	Xq2.7
Gene	FMR-1	Pi	CFTR	dystrophin	Factor VIII	Factor IX
Length (kb)	>80	10	250	2400	187	34
Number of exons	—	5	27	79	27	8
mRNA (nucleotides)	4800	1434	6500	14000	7053	1383
Mutations						
Spontaneous mutation frequency	—	—	—	1.10^{-4}	5.10^{-5}	3.10^{-6}
Type						
insertion/deletion (large)	—	5%	1%	66%	5%	10%
Point mutation	1%	95%	99%	—	95%	90%
VNTR	99%	—	—	—	—	—
Spontaneous	—	—	—	33%	50%	—
Number of different mutations	—	>75	>400	>300	>100	>115
Prevalent mutations	—	Z	δF508	—	—	—

[a]AAT, α_1-antitrypsin; CF, cystic fibrosis; DMD, Duchenne muscular dystrophy.

To assess individual DNA sequence variation on a total genome scale the parallel processing approach is more efficient. This strategy seeks to analyse as many fragments as possible in parallel rather than serially. It can maximize the information content of the analysis of a single genome by simultaneously targeting many loci. For the analysis of genomes of higher organisms hybridization or amplification techniques using different types of repetitive sequences have to be applied to visualize selectively particular subsets of fragments. In this way it is possible to analyse comparatively such genomes at many sites in different individuals. The individual differences observed can be classified according to the probe and/or primer used. Such a classification, as will be discussed in Chapter 2, can be extremely useful in combining the data on individual differences (for example, in pedigrees, healthy and diseased individuals, tumour vs normal) to datasets comprising both physical and genetic mapping information on the genome of the species of interest.

One way of achieving comprehensive parallel analysis of a genome is by 2D gel electrophoretic formats based on the combination of different separation criteria of DNA fragments. By combining efficient electrophoretic mutation scanning techniques in a two-dimensional display of fragments, the information content of a single analysis can be very high. Complete genomes of lower organisms can be resolved in this way, while simultaneously allowing comprehensive detection of sequence variations at any given site in the genome. A particular two-dimensional electrophoretic scanning system, 2D DNA typing, is the subject of this volume. Some basic aspects of two-dimensional DNA electrophoretic systems will be discussed below.

1.4   TWO-DIMENSIONAL SEPARATION OF DNA FRAGMENTS

1.4.1   Introduction

In the previous sections several electrophoretic techniques have been discussed which can be used in combination to constitute a two-dimensional separation format for DNA molecules. Two-dimensional separation techniques can be based on a combination of similar or completely independent separation criteria (Fig. 1.6). In the first case, fragments are clustered along a straight diagonal after two-dimensional electrophoresis (Fig. 1.6a). The use of a second restriction enzyme to digest size-separated fragments generated by digestion with a first restriction enzyme will result in only the area below the diagonal being used to resolve the fragments (Fig. 1.6b and c). If completely independent criteria are combined substantial deviations from the diagonal will arise after separation.

Initially, 2D DNA separation methods were applied to resolve enzymatic digests of relatively simple prokaryotic genomes. Usually all fragments could be resolved in agarose or polyacrylamide gels and visualized by ethidium bromide staining or end labelling. For more complex molecules only selective visualization of sets of fragments from the genome of interest will allow interpretable analysis. This can be accomplished by hybridization analysis using probes targeting multiple sites in the genome of interest, or by selective end-labelling of separated fragments. Alternatively, PCR methods can be applied to amplify selectively particular DNA sequences such as those bordered by repetitive sequences (interrepeat sequences) or multiple fragments of a contiguous stretch of DNA, such as a complete gene.

1.4.2   Enzymatic digestion as second dimension criterion

Early applications of two-dimensional separation of DNA fragments involved the use of two subsequent enzymatic treatments of genomic DNA either with two restriction endonucleases or with a restriction enzyme and S_1 nuclease (reviewed in Yee and Inouye, 1984). In general, the procedure involves restriction enzyme digestion and a first-dimensional size separation, followed by the incubation of a piece of polyacrylamide or agarose gel in a solution, containing the second enzyme, to further digest the DNA fragments. In view of the use of similar size separations in both dimensions, only the area below the diagonal of size-separated fragments can be used for resolution of fragments (Fig.1.6b).

1.4.2.1 S_1 nuclease digestion

Two-dimensional electrophoresis involving S_1 nuclease (Yee and Inouye, 1984) can be used to detect heteroduplex molecules in complex populations of homoduplex molecules. The method includes digestion of genomic DNA with a restriction enzyme recognizing four bases, followed by a denaturation/renaturation step, separation of the fragments by size in a neutral polyacrylamide gel, incubation of the first dimension lane in a solution containing S_1 nuclease and, finally, a second dimension electrophoresis in a second neutral polyacrylamide gel. As a result only few spots, derived from those fragments which are deviating from the wild type, can be observed which all occur below the expected diagonal of size-separated fragments. These fragments correspond to the smaller-sized fragments resulting from the presence of a stretch of mismatched basepairs (see also the previous section on S_1 nuclease reactivity). This technique of two-dimensional DNA separation can be used for scanning small fragments for sequence variation but is much less applicable to problems relating to genome scanning. In particular, the denaturation/renaturation step is not possible in the analysis of complex genomes. This is due to the presence of abundant repetitive sequences leading to the formation of aberrant heteroduplexes.

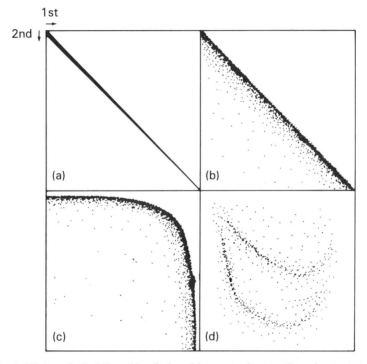

Fig. 1.6. Schematic depiction of the display of fragments after two-dimensional gel electrophoretic separations of macromolecules based on identical or independent criteria in the first and second dimension separation. (a) Identical first and second dimension separation criteria. (b) Separation criteria are identical in first and second dimensions, but fragments have been digested (e.g. with a restriction enzyme) between the two separations. (c) Separation criteria are identical in first and second dimensions and fragments have been digested between separations. The gel matrix in the first dimension is different from the one in the second dimension (e.g. agarose vs PAA or different percentages of PAA). (d) First and second dimension separation criteria are independent.

1.4.2.2 Digestion with a second restriction enzyme

When two consecutive restriction enzyme digestions are applied a two-dimensional display of fragments will occur similar to the one described above for S_1 nuclease. That is, all fragments are below the diagonal (Fig. 1.6b). Bands arising in the first dimension separation can be individually excised and treated with the second enzyme. This approach has been successfully applied to analyse and isolate repetitive sequences in *Drosophila* DNA (Gilroy and Thomas, 1983). Alternatively, the whole 1D gel lane itself or its separation pattern transferred to a DEAE cellulose membrane can be incubated with the second enzyme.

For the visualization of the fragments resolved on a 2D gel three routes can be pursued. First, DNA dyes such as ethidium bromide can be applied to visualize all fragments on the gel. This has been applied to the analysis of simple prokaryotic genomes such as *Escherichia coli* and *Bacillus subtilis* and the mycoplasma *Acheloplasma laidlawii* to analyse repeated DNA sequence families and to estimate genome sizes (Potter *et al.,* 1977; Yee and Inouye, 1982; Boehm and Drahovsky, 1984; Poddar and Maniloff, 1986). With appropriate restriction enzyme combinations 180 to 1312 restriction fragments varying in size between 0.350 and 5 kb were observed allowing accurate estimates of the genome size of a variety of prokaryotic species.

Second, transfer of the separation pattern to a membrane and hybridization analysis using specific probes allows the visualization of particular subsets of fragments. When applied to a more complex genome, such as that of the protozoan *Tetrahymena thermophilae* which is roughly 200×10^6 bp, hybridization analysis was used to visualize a repetitive subset of fragments (Hüvos *et al.,* 1988). This approach allowed the resolution of 80 members of a repeat family and was capable of detecting interstrain differences in this organism. Similar types of analyses have been performed with *Saccharomyces* (Rogan *et al.,* 1991), human genomic DNA (Sakaki *et al.,* 1983), and mouse genomic DNA (see below). Polymorphisms were detected in gene families such as the major histocompatibility complex (MHC) of the mouse (Sheppard *et al.,* 1991) and intracisternal A particles (Au *et al.,* 1990; Sheppard *et al.,* 1991) and included the detection of methylation polymorphism (Fanning *et al.,* 1985). In spite of the fact that only half of the gel area could be used for resolution of the fragments, still up to 370 spots could be detected using this approach. The use of agarose gels in these separation methods resulting in rather faint and blurry spots limits the resolution (e.g. Sheppard *et al.,* 1991). This can, however, be compensated for by using polyacrylamide gels and electroblotting protocols (Uitterlinden *et al.,* 1989a).

A third way of visualizing particular subsets of fragments is by selective end labelling or T4 polymerase strand displacement labelling of restriction fragments (Sambrook *et al.,* 1989). That is, after digestion of genomic DNA with a first restriction enzyme, fragments are labelled and subsequently further digested with a second, more frequently cutting, enzyme. This has been applied in the analysis of strains of *Escherichia coli* (Yi *et al.,* 1990) and in the analysis of the *Drosophila* and the mouse genomes (Hatada *et al.,* 1991). The latter method, restriction landmark genomic scanning (RLGS), allows up to 2000 spots to be analysed in a single gel and was shown to be capable of detecting DNA RFLP polymorphisms among different individual mouse DNA samples. The technique is based on the two-dimensional separation of end-labelled DNA restriction fragments in which the first dimension is run in agarose gels and the second dimension in PAA (polyacrylamide) gels. By using two different matrices for the separations, efficient use is made of the 30 cm × 45 cm 2D gel area resulting in up to a few thousand spots per pattern (Fig. 1.6c).

The application of RLGS for mouse genome mapping is based on the analysis of 26 DBA × C57/Bl6 recombinant inbred strains and inter-species backcrosses of C57/Bl6 × *Mus spretus*. Between C57 and DBA 13.4% (51/378) variant spots were detected; between *M. musculus* and *M. spretus* this was 51% (188 out of 369). This illustrates the major drawback of the method, i.e. the very low level of polymorphism that can be detected. As a result it cannot be used for human genetic analysis. Indeed, when two unrelated (Japanese) individuals were compared by RLGS only 0.15% spot variants were detected (Hayashizaki, personal communication). This is explained by the low informativeness of restriction enzyme recognition sites in general. The technique has been applied to the analysis of some tumours to detect amplifications (in a breast tumour 10 spot variants showing 4–7 times amplification were detected). The most commonly detected genetic change observed in tumours, i.e. loss of chromosomal fragments (indicated by the loss of a spot variant), is inherently difficult to detect by RLGS because intensity decreases of two-fold must be detected. Indeed, reliable detection becomes virtually impossible when the tumour samples to be analysed are contaminated with normal cells, as is often the case.

1.4.2.3 Pulsed field electrophoretic separation

The principle of two consecutive restriction enzyme digestions as basis for a two-dimensional DNA separation can be extended to very infrequently cutting enzymes. After digestion of genomic DNA with such an enzyme (for example Not I), separation of the resulting fragments by pulsed field gel electrophoresis (PFGE; Schwartz and Cantor, 1984) or field inversion gel electrophoresis (FIGE; Carle *et al.,* 1986) in the first dimension is followed by *in situ* digestion with a more frequently cutting enzyme and size separation in the second dimension in normal agarose gels. This approach was applied in the analysis of multigene families such as the murine T-cell receptor complex (Woolf *et al.,* 1988), the Ly-6 multigene family (Kamiura *et al.,* 1992) and the human immunoglobulin heavy chain variable region (Walter and Cox, 1989). In addition, first dimension separation of partial digests by FIGE, followed by *in situ* complete digestion of the fragments and FIGE separation in the second dimension, has been used to create macrorestriction maps of microorganisms such as *Mycoplasma mobile* (Bautsch, 1988). Although the method is suitable for large-scale mapping of genomic regions or even entire genomes of microorganisms, the effectiveness as a genome scanning method is quite low in view of the low number of fragments obtained and low mutation scanning efficiency. That is, only large insertions or deletions and variation in the recognition site of the restriction enzyme can be detected by this approach.

In summary, restriction enzyme digestion is helpful by generating many fragments to be analysed. However, in the absence of an independent second dimension separation criterion, only the recognition sites of the restriction enzymes used are screened for variation (see Table 1.2). In general, only the area below the diagonal in a 2D separation pattern is used in these 2D formats. The actual gel area used can be increased by taking different gel media or concentrations of gel media for the subsequent separations (e.g. Hatada *et al.,* 1991; see also Fig. 1.6c).

1.4.3   Conformation as second dimension separation criterion

Attempts to combine two independent electrophoretic separation criteria involved the separation on the basis of conformation of DNA molecules under different circumstances such as the presence of intercalating dyes (Meyer and Hildebrandt, 1986; VanWye *et al.,* 1991) and the use of different electric field strengths, buffers and matrices (Bell and Byers, 1983; Shinomiya and Ina, 1991; VanWye *et al.,* 1991). Such methods can reveal differences in GC content, DNA bending, single strand conformation etc., which can be applied for mutation detection when used in combination with SSCA (Kovar *et al.,* 1991). Although they can reveal minute differences, the resolution obtained is low in that only few spots will deviate from the diagonal. These methods therefore do not allow large-scale scanning of genomic regions.

1.4.4   DGGE in the second dimension

The denaturing gradient principle as applied in DGGE was used in an early stage of development as a second independent separation criterion in two-dimensional electrophoretic analysis of highly complex mixtures of hundreds of different restriction fragments (Fischer and Lerman, 1979a,b; see also previous section on DGGE). Since the separation criteria are now truly independent, optimal use can be made of the gel area (Fig. 1.6d). The method could resolve essentially all Eco RI restriction fragments of the *Escherichia coli* genome (estimated size: 4399 kb) and detect a 50 kb insertion of lambda DNA in the *E. coli* genome (Fischer and Lerman, 1979a). In addition, the 2D method has been used to resolve digests of complete genomes of several other microorganisms with sizes ranging from 724 to 1833 kb (Poddar and Maniloff, 1989).

In all the 2D systems so far discussed, migration is to a very large extent dependent on fragment size in both first and second dimensions. One particularity of the DGGE separations, however, is their virtual independence of the length of a fragment owing to different migratory behaviour in PAA gels of partially melted DNA molecules. Initially the PAA gels will sieve the fragments according to their size but, on reaching the $T_m(50)$ of the fragments, this sieving is in most instances overridden by the sieving according to the partially single-stranded structures. As a result fragments of different size but sharing a particular melting domain will migrate to similar positions in the DGGE dimension after 2D separation. For sheared total genomic DNA, for example, this will result in the appearance of streaks in a horizontal direction at different levels in the second dimension gel pattern. This results in a two-dimensional display of melting domains present in a given DNA molecule, which can also be used for comparative purposes. This feature has been exploited in the separation of sheared pBR322 fragments (Fischer and Lerman, 1983; Lerman *et al.*, 1981, 1984).

Deviations from the normal position in a 2D gel can occur as a result of changes in length or sequence content of the fragment. The sensitivity in this respect is determined by the percentage polyacrylamide (PAA) and the gradient of denaturants, respectively, and by the size of the gel. Normally, 6% PAA gels are used which allows for the separation of fragments of 0.1 to roughly 2 kb with a resolution of approximately 5% of the fragment length using standard size equipment (see also Chapter 3). Higher percentages of PAA will allow smaller length differences to be detected when present in fragments of a few hundred basepairs. For standard DGGE separations a gradient of 0–80% denaturants is usually applied. This will allow most fragments to reach their melting point and to focus in the gel. The sensitivity for detection of sequence variations is lower in this gradient but can be increased by taking gradients spanning a shorter range of denaturants.

The efficient use of the 2D gel area (Fig.1.6d) and the high sensitivity for sequence variations, such as point mutations, independent of the restriction enzyme used (see Table 1.2), make 2D DNA separation with DGGE in either the first or the second dimension, the ideal method for analysing complex genomes. Hence, DGGE was selected as a separation criterion in two-dimensional DNA typing, the subject of this book. Before discussing the place of 2D DNA typing in the field of genome analysis (see Chapter 2) and its basic methodology (see Chapter 3) in the next sections some general characteristics of the system will be discussed.

1.4.5 Two-dimensional DNA typing

The efficient use of the 2D gel area and the high sensitivity for sequence variations, independent of the restriction enzyme used, make 2D DNA separation with DGGE in either the first or the second dimension, the ideal method for analysing complex genomes by so-called 'genome scanning' (Vijg and Uitterlinden, 1987; Uitterlinden *et al.*, 1989a). Genome scanning involves the analysis of many sites in the genome simultaneously for sequence variation among individuals, with the ultimate aim of correlating the variants observed with phenotypic characteristics.

2D DNA typing allows the efficient detection of deviations from a normal sequence at many possible sites in the genome of interest. It does not, however, provide information on the location in the genome of the variation. In principle, a two-dimensional DNA separation pattern represents a display of fragments derived from a contiguous sequence of basepairs. The relation of the two-dimensional display of the fragments to their original sequence in the genome of origin is usually not known *a priori*.

For small genomes, yielding few restriction enzyme fragments, the relationship between gel position and genomic position can be easily determined by partial

digestion experiments. For larger regions or genomes, however, additional characterization is necessary to obtain interpretable 2D separation patterns. A logical
next step in this respect is hybridization with probes. Probes detecting overlap
of certain restriction sites in the genome of interest will identify adjacent restriction fragments and allow for the construction of contiguous sequences (contigs)
corresponding to particular spots in the seperation pattern. When it is known
that all possible fragments from a given genome are resolved in a single 2D
gel and that no spots are derived from more than one restriction fragment this
approach will allow for the construction of a physical '2D' map of a genome.
Alternatively, such a physical map can be composed by subsequent hybridization
analysis with multiple short oligonucleotides (Lehrach *et al.,* 1990). In analogy
to this mapping approach, even the sequence of the fragments might also be
obtained by means of hybridization with multiple overlapping oligonucleotides
(Drmanac *et al.,* 1989). For larger genomes such a 2D genome mapping approach
becomes unfeasible owing to the great number of fragments; multiple fragments
are likely to be present at a certain spot position. Of course for smaller regions
contig mapping is possible using unique sequences as probe derived from cosmids or YACs.

The most efficient method of extracting comprehensive information from a
genome by 2D DNA typing is by using probes for repetitive sequences. When
repetitive sequences are used as probes in hybridization analysis, multiple sites
in the genome can be analysed simultaneously, but information on order of fragments is lost and physical mapping information can no longer be retrieved from
the 2D pattern. However, each spot can be assigned a third characteristic (i.e.
partial sequence content, apart from its length and $T_m(50)$) which makes it part
of a group of related sequences with a particular function, organization and/or
location on chromosomes. Such repetitive sequences can include 'anonymous'
repeats such as micro- and minisatellites, retroviral elements occurring in many
genomes but may also include motifs specific for groups of protein encoding
sequences such as zinc-finger motifs, homeobox motifs, etc. A particular form
of 2-D DNA typing in which repeats are used but in which the physical and
genetic mapping information can be preserved and exploited is the use of primers
for repetitive elements in (PCR) amplification reactions. Interrepeat regions can
be selectively amplified from sources with a known physical map location such
as cosmids, YACs and somatic cell hybrids and used as probes in comparative
studies of different individuals. This allows combined physical and genetic mapping of a particular region (Uitterlinden *et al.,* 1991a).

1.5 SUMMARY AND DISCUSSION: GENERAL CHARACTERISTICS OF 2D DNA TYPING

From the previous pages the picture has emerged of a form of genetics in which
the DNA of the genome has become amenable to direct analysis rather than one
in which it is necessary to rely on phenotypically expressed traits. The avail-

ability of new advanced DNA analytical and manipulative technology justifies the prediction that, even for some higher organisms with large and complex genomes including humans, all the genes will be isolated and sequenced by the year 2005. Such an achievement will certainly form the solid basis for many future studies in biology and medicine alike. It will, however, not solve the problem of genetic variation. In view of its fluid character (new mutations will continue to occur) genetic variation needs to have methods available for monitoring the DNA sequences of entire genomes on a permanent basis.

When reviewing the methods available to identify DNA sequence variation, DGGE appears to be the method of choice, allowing DNA fragments to be effectively scanned for any given type of sequence variation (see Table 1.2). For total genome scanning, a parallel approach combining DGGE with size separation after restriction enzyme digestion (or the amplification of multiple fragments) might have distinct advantages over a serial approach consisting of repetitive determinations using robotics.

Two-dimensional DNA typing has its major application in comparative studies aimed at rapidly genotyping individuals. A few such studies will be discussed in Chapter 4. As described above, this mainly concerns the use of probes for repetitive sequences in genomic DNA. Apart from the analysis of genomic DNA it is also possible to analyse only the protein-encoding part of the genome by analysing cDNA prepared from mRNA populations. In view of the complexity of mRNA populations in cells, 2D DNA typing can offer a high resolving capacity to analyse many individual mRNA species in parallel from a given cell type (see Chapter 5). For comparative purposes such cDNA libraries can be prepared from different cell types and from different individuals. The introduction of amplification primers and application of probes (such as the aforementioned functional motifs) allows particular subsets to be analysed. In this respect, two-dimensional DNA typing could allow the simultaneous analysis of many candidate genes for particular disorders.

Finally, 2D DNA typing can help in the comprehensive analysis of large regions in the genome. High resolution mutation analysis of especially the larger genes is at present impossible but might become feasible by 2D DNA typing (see Chapter 5).

So far, 2D DNA typing has mainly been focused on the use of anonymous repeats (micro- and minisatellites) as probes in studies of genomic instability and in association/linkage studies. Before some of these applications are treated in detail in Chapter 4, the basic principles and strategies of current genome analysis projects will be discussed.

2

Genome analysis: general considerations

2.1 INTRODUCTION

In the last few years genome analysis has drawn ample attention in the light of the unprecedented self-imposed task of mapping and sequencing the entire human genome within 15 years. In the framework of this, genomes of several other lower organisms are already in the process of being mapped and sequenced to provide systems in which to test methods of genome analysis. Not only will these latter projects provide insight into the biology of the organisms themselves but also, by comparative analysis, into that of a host of other organisms including the human species.

We will not attempt to discuss the rationale of such a momentous task in terms of whether or not this is the most efficient way of advancing the science of biology and medicine; genome projects for human, mouse, yeast, nematode and *E. coli* have already been initiated. Therefore a discussion of new technology for genome analysis simply has to conform to the realistic situation of a massive preoccupation during this and the next decade with charting genomes.

Genome projects will inevitably result in the location of genes for medical conditions and other human traits and for economically important traits in animals and plants. Depending on the strategy followed, either the exhaustive sequencing of all DNA or mapping the entire genome and then sequencing only those parts that encode genes, the molecular blueprints that underlie species-specific characteristics will sooner or later become available in the form of large databases.

In this chapter we will provide an overview of the general considerations that underlie genome analysis, the different methods that are being used and the imminent and long-term results to be expected. Attention is focused on the various possibilities to identify genes by associating particular phenotypes with the underlying genetic variation, with emphasis on the possible role to be played by two-dimensional DNA typing.

2.2   MAPPING GENOMES

2.2.1   Physical mapping

The ultimate aim in the analysis of a genome of interest is the delineation of its basepair sequence. This will allow the identification of genes and their regulatory elements, and provide an important basis to gain more insight into the biology of the organism under study. In most prokaryotes, which have small genomes, genes are continuous (there are no introns) and tightly packed with sometimes one DNA sequence encoding more than one protein. Genes in genomes of higher organisms tend to be dispersed and are discontinuous, i.e. organized in non-coding introns and protein coding exons. Distances between exons can be large (up to tens of kilobasepairs) and coding regions can be separated by enormous stretches of 'junk' DNA.

For very small genomes of up to several thousand basepairs the entire basepair sequence can effectively be determined by consecutive DNA sequencing reactions performed for contiguous stretches of DNA (see Chapter 1). For larger genomes this process has to proceed in different steps, from the identification of chromosomes and other defined genomic regions down to the ordering of particular fragments along the sequence and, finally, the determination of the actual basepair sequence itself. This process of ordering the DNA of the genome down to the basepair sequence is usually referred to as physical mapping. Because of the large size of most genomes the majority of physical mapping techniques first involve the generation of fragments which are much smaller than the original DNA molecule and fall within the range of resolution of electrophoretic separation techniques. Fragment order can then be derived directly from electrophoretic separation patterns. More often, however, one uses cloned copies of fragments of genomic DNA in the form of recombinant DNA libraries. Libraries that are most commonly used in this respect are cosmid libraries which can harbour up to 45 kb of DNA. On the basis of overlap between inserts of such clones one can eventually determine the order of the fragments in the genome of origin. To determine overlap within libraries currently two different approaches are taken, based on serial and parallel analysis. Serial analysis involves the restriction mapping of one or a few clones per analysis (Coulson and Sulston, 1988; Carrano *et al.,* 1989), while the parallel approach is based on consecutive hybridization rounds, using small oligonucleotides, of complete ordered libraries consisting of up to ten thousand clones (see below; Lehrach *et al.,* 1990).

In the early days of recombinant DNA technology either very large fragments (e.g. chromosomes) or very small fragments (e.g. restriction fragments) could be analysed. That is, cytogenetic analysis allowed the identification of chromosomes and chromosomal regions with an informational size range of tens of millions of basepairs or higher. On the other end, without anything in between, restriction fragment mapping had a size range of only thousands of basepairs. This 'size-informational gap' posed considerable logistical problems for the analysis of larger genomes, such as that from humans, simply because of the very large number of clones or fragments to be analysed.

More recently, cloning methods and electrophoretic separation techniques have been developed which can handle much larger fragments, thereby bridging the gap between cytogenetics and restriction mapping. These methods include the electrophoretic separation of very large restriction fragments of up to several million basepairs through pulsed field gel electrophoresis (PFGE; Schwartz and Cantor, 1984), modified by Carle and Olson (1984, 1985) and Smith and Cantor (1986) or by field inversion gel electrophoresis (FIGE; Carle *et al.,* 1986). In addition, the use of rare-cutter linking/jumping libraries allows the bridging of distances up to several hundreds of kilobasepairs, separating recognition sites of enzymes such as Not I (Poustka and Lehrach, 1986). Furthermore, the cloning of very large fragments in yeast artificial chromosomes (YACs; Burke *et al.,* 1987) enables very large regions of DNA (up to around one million basepairs) to be physically isolated, allowing the construction of libraries containing complete genomes in a relatively small number of clones. More recently, radiation hybrid mapping (Cox, 1991) enables one to make collections of somatic cell hybrid clones with a very dense spacing of radiation-induced breakpoints once every few million basepairs along a chromosome (region) of interest. All these methods provide the means to establish the order of, first, small fragments within the larger ones and then the larger ones within cytogenetically distinguished regions of a complete genome.

One of the first species for which a complete physical map of its genome has been made is *Escherichia coli* (Smith *et al.,* 1987). The map consists of 22 Not I restriction fragments, ordered along the single circular chromosome of 4.7 Mb as determined by PFGE. For the human genome, hybridization analysis of PFGE gel separation patterns allowed several regions to be physically mapped, such as the major histocompatibility complex region in humans (the HLA complex) spanning over 3 Mb (Lawrance *et al.,* 1987) and regions around disease genes such as the Duchenne muscular dystrophy region on Xp21 (Kenwrick *et al.,* 1987; Burmeister and Lehrach, 1986). In the human genome project one is now collectively working to create complete physical maps of all the chromosomes. The smallest one, chromosome 21, has already been completely physically mapped as an ordered array of YACs (Chumakov *et al.,* 1992). A physical map of the complete human genome is being established by fingerprinting of random YACs which now has a coverage of approximately 20% (Bellanné-Chantelot *et al.,* 1992).

In physical mapping genomic fragments are ordered in a linear array on the basis of consensus sequences. In this respect its final step is necessarily the determination of the complete nucleotide sequence of an entire genome, that is, a consensus genome. Physical mapping, therefore, does not account for individual differences in DNA sequence. By contrast, in genetic mapping the detection of individual DNA sequence variation is of central importance.

2.2.2 Genetic mapping

A genetic map of an organism is based on the observation of variant genotypes in populations of that species, reflecting DNA sequence variations at particular (marker) loci. A genetic map then displays the genetic distance between polymorphic marker loci distributed over the chromosomes. The distances between such marker loci on a chromosome are estimated from segregation studies in pedigrees. If two marker loci are located close together, their allelic variants are likely to be co-inherited and the two loci are said to be linked. In that case the chance that a recombination event between homologous chromosomes (crossing-over) will separate the two variants is small. The frequency of recombination, however, will increase with the distance between loci. Genetic distances are usually expressed in centiMorgans (cM) whereby loci said to be 1 cM apart show recombination in 1% of the meioses. For the human genome 1 cM genetic distance roughly corresponds to an average physical distance of 1×10^6 bp, but over five-fold deviations from this general rule have been observed owing to local variations in recombination frequency. The relationship between physical and genetic distance is also species dependent. For example, yeast has a much higher frequency of meiotic recombination than mammalian organisms whereby during yeast meiosis, the ratio of genetic to physical distance is only 1 cM per 2–3 kb (Kaback et al., 1989). Moreover, since the basis for a genetic map is the scoring of meiotic recombination events between (marker) loci occurring in sperm cells and oocytes, genetic maps can be constructed for males and females, separately. A peculiarity in this respect is that recombination has been shown to be more frequent in females than in males, which results in genetic distances in general being shorter on male genetic maps. In view of the above it should be concluded that a genetic map is more a reflection of local activities of recombination enzymes than a display of DNA sequence organization. Thus, like physical mapping, genetic mapping provides information on the order of marker loci on contiguous fragments but on a very different basis (for reviews, see Lalouel et al., 1986; Lander and Botstein, 1986; Lander, 1988).

Analysis of genetic polymorphisms has been recognized for a long time as a means of genetic mapping, i.e. ordering particular genetic traits, such as heritable diseases, as genes along a linear genome. However, only after the discovery of DNA polymorphisms (RFLPs) did a vast supply of marker loci become available to create genetic maps (Kan and Dozy, 1978). It was realized that in fact any given genetic trait can be localized based on the creation of a genetic map of the human genome consisting of DNA polymorphisms distributed over all chromosomes (Botstein et al., 1980). The collections of RFLPs being compiled over the years has led to the creation of genetic linkage maps of chromosomes (White et al., 1985), even to an extent where a complete genetic map of the human genome was claimed, albeit with a rather uneven distribution of markers

(Donis-Keller *et al.*, 1987). At present, genetic maps have been compiled for most human chromosomes.

More recent genetic maps consist entirely of $(GT)_n$ loci (see below), such as for mice (Dietrich *et al.*, 1992) and humans (Weissenbach *et al.*, 1992). When combining the different markers on these maps an ever-increasing genetic resolution can be obtained allowing any given particular gene to be mapped in the proximity of one or more of these genetic marker loci. To date there are approximately 1000 human $(GT)_n$ markers giving an average resolution of more than 1 marker per 5 cM. To be able to extract the necessary genetic information from particular families it is necessary to have a highly informative marker every 1–2 cM. It is expected that over 4000 human polymorphic $(GT)_n$ markers will become available within the next 2–3 years (Todd, 1992).

These collections of DNA polymorphisms have been successfully applied in the identification and isolation of disease genes by pedigree analysis using the genetic linkage approach. This strategy, which has been termed 'reverse genetics' (Orkin, 1986), 'forward genetics', and is now more accurately described as 'positional cloning' (Collins, 1992), has enabled the isolation of the affected gene and the identification of the disease-causing mutation(s). Thereby, the causes of some of the most common human monogenic diseases have been elucidated, e.g. Duchenne muscular dystrophy (DMD; Monaco *et al.*, 1986; Koenig *et al.*, 1987; Burghess *et al.*, 1987) and cystic fibrosis (CF; Riordan *et al.*, 1989; Rommens *et al.*, 1989).

A first recognition of the chromosomal area containing the disease gene of interest is often provided by the availability of patients having cytogenetic abnormalities such as chromosomal deletions and translocations. This was the case in cloning the genes for DMD, retinoblastoma, and neurofibromatosis type 1. After this unprecise localization, genetic linkage analysis can provide a more detailed localization using DNA markers in the identified region. For some monogenic diseases, such as cystic fibrosis and Huntington's disease (HD), however, such chromosomal abnormalities were not found in patients and this has made the final identification of the gene very troublesome (Roberts, 1990a; Morell, 1993; see also 2.4).

With increasing numbers of DNA markers more evenly distributed over the chromosomes genetic linkage studies become more accurate in localizing the gene of interest. For simple genetic disorders gene identification by positional cloning has therefore become a rather straightforward exercise, provided that the disease phenotype can be unambiguously recognized and sufficient informative pedigree material is available.

Once genetic linkage has been firmly established the area of the genome defined by genetic markers flanking the disease locus must be scrutinized for the presence of candidate genes. This involves the generation of a physical map of the region by ordering increasingly smaller fragments along the chromosome, going from PFGE maps and YAC clones spanning the region to cosmid clones. Important clues as to where active genes are present can come from a search for HTF islands (GC-rich stretches which usually are located 5' of a gene) and from comparative analysis with other species to look for conserved and expressed sequences. The final identification of the gene responsible for the disease under study is dependent on the finding of a DNA sequence variant occurring in patients and not in normal controls. It should be noted that even when linkage with defined marker loci can be found, the identification of the actual disease gene can be very difficult. For Huntington's disease, for example, this took 10 years, while in the case of cystic fibrosis the gene was identified only after linkage disequilibrium narrowed the region in which it resides.

2.2.3 DNA markers

The identification and screening of many marker loci started with the introduction of Southern blot analysis (Southern, 1975) which led to the discovery of so-called restriction fragment length polymorphisms (RFLPs). RFLPs were first detected for the globin gene locus (Kan and Dozy, 1978; Jeffreys, 1979). Soon thereafter RFLPs were described for so-called anonymous DNA segments (Wyman and White, 1980). Especially the latter type of sequence displays DNA polymorphisms, since as non-coding DNA it is under much less evolutionary selective pressure to preserve its sequence in populations (see also below).

The first RFLPs discovered were based on the absence or presence of a restriction enzyme recognition site and can be referred to as restriction site polymorphisms or RSPs. In Fig. 2.1 an example is shown of an Xba I RSP at the human D21S16 locus. The enzyme dependency of the detection of such RFLPs is shown in the same figure by digesting the same DNAs with Rsa I which fails to detect an RSP in the same sequence. Extensive searches for such RSPs in the human genome showed that on average 1 in 300 to 1 in 1000 basepairs is polymorphic in the population for autosomes and the X chromosome, respectively (Cooper et al., 1985; Hofker et al., 1986). In this respect CpG sites show a higher frequency of polymorphism owing to the susceptibility of methylated cytosine to undergo mutation (Barker et al., 1984).

The frequency of polymorphic basepairs in human DNA measured through RFLP analysis is likely to be an underestimation since it is a very insensitive way of detecting point mutations. Several other methods are now available, such as DGGE analysis, which can detect sequence variation in any given stretch of basepairs much more efficiently (see Chapter 1). These methods have only recently been introduced to exploit DNA sequence variation in genetic linkage analysis of disease. A particular advantage of methods such as DGGE is that they allow multiple alleles of a given basepair sequence to be discerned as a result of the presence of several polymorphic basepairs in the sequence (e.g. Sheffield et al., 1992b; see also Fig. 1.2 in Chapter 1). Such multi-allelic systems will increase the informativeness of the locus in genetic analysis.

Another set of multi-allelic markers is based on a completely different type of polymorphism, involving repetitive sequences, termed micro- and minisatellites. Loci with such sequences are found in multiple (several to hundreds of thousands) copies dispersed throughout the genome of many different species.

In this respect, they represent ideal candidates for genome scanning studies because a single probe, corresponding to a core sequence shared by a set of micro- and/or minisatellite loci, can detect many genetic marker loci simultaneously. In Table 2.1 an overview is presented of the different forms of repeat sequence polymorphisms found in the human genome. In view of their importance in 2D DNA typing, we shall first discuss the micro- and minisatellite sequences in more detail. However, other repetitive sequences can also be exploited for 2D DNA typing. These include the dispersed elements such as Alu I which will be discussed later on.

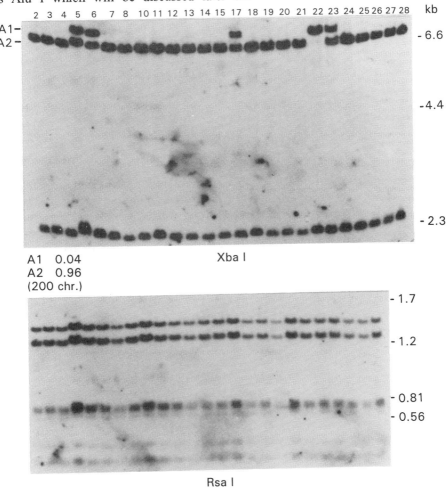

Fig. 2.1. RFLP detection at the D21S16 locus by Southern blot hybridization analysis. Genomic DNA was digested with either Xba I or Rsa I and hybridized with a probe detecting the locus. Only Xba I reveals the presence of an RSP which results in detection of two alleles (A_1 and A_2 of which A_1 is very rare) while Rsa I resulted in a monomorphic banding pattern among the 25 unrelated individuals tested.

Table 2.1. Human repeat sequence polymorphisms

Type of repeat	Distribution	Copy number per haploid genome	Type of polymorphism	Polymorphic fraction[a]
Alu I	Dispersed	0.5×10^6	Insertion/deletion	Infrequent
			Point mutation	50%
			PolyA-VNTR	60%
Kpn I	Dispersed	$1 \times 10^3 - 1 \times 10^5$	Insertion/deletion	Infrequent
			Point mutation	?
			PolyA-VNTR	?
Alphoid satellite	Centromeric	1×10^4	VNTR	?
Minisatellite	Dispersed/ subtelomeric	2×10^4	VNTR	50–75%
			Point mutation	?
Microsatellite	Dispersed/ clustered	$1 \times 10^3 - 1 \times 10^5$	VNTR	50–75%
			Point mutation	?
LTR	Dispersed	5×10^3	Insertion/deletion	Infrequent
			Point mutation	?

[a]Estimated percentage of copies with a polymorphism.

2.2.4   Micro-and minisatellite sequences

Micro- and minisatellite loci consist of short sequence units which are tandemly repeated. Some representatives of these repeat loci show polymorphic variation in the number of repeat units at the locus and can be termed variable number of tandem repeat or VNTR loci. Fig. 2.2 shows a schematic representation of the structure of minisatellites and VNTR polymorphism. Microsatellite VNTRs are referred to as microsatellites, simple sequence repeat motifs (SSRMs), short tandem repeats (STRs), variable number of dinucleotide repeats (VNDRs) or simple sequence length polymorphisms (SSLPs).

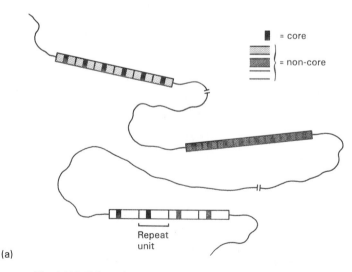

(a)

Fig. 2.2(a). Schematic representation of minisatellite sequences.

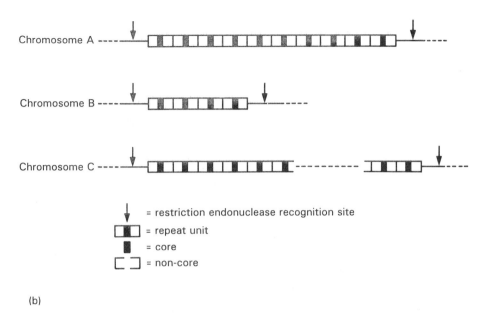

(b)

Fig. 2.2(b). Schematic representation of minisatellite VNTR type polymorphism.

2.2.4.1 Minisatellites

Some VNTRs have been shown to be extremely polymorphic with more than a hundred alleles in the population. A striking example of such VNTR polymorphism is shown in Fig. 2.3. Hybridization analysis of 25 different human individuals at the D1S7 VNTR locus resulted in the detection of close to 50 different alleles of this locus. Notably, minisatellite repeats were originally described as unrelated regions of short tandem repeats in anonymous DNA regions (Wyman and White, 1980) or in the proximity of genes such as the α-globin gene (Higgs *et al.*, 1981; Jarman *et al.*, 1986), the insulin gene (Bell *et al.*, 1982; Ullrich *et al.*, 1982), the ζ-globin gene (Prondfoot *et al.*, 1982) and the cHa-ras proto-oncogene (Capon *et al.*, 1983). Later, Jeffreys *et al.* (1985a) organized them into sets of repeat loci (termed minisatellites) on the basis of common short stretches of sequence homology, the so-called core sequences. When such a core sequence is used as a probe in Southern blot hybridization under relaxed hybridization and washing conditions many minisatellites are detected simultaneously. This is illustrated by Fig. 2.4 in which core probe 33.15 and locus probe pλG3 are used to hybridize the same human genomic DNA separation patterns. In this example of a human paternity test, the bands found with the locus probe also occur in the patterns obtained with the core probe.

The relative importance of the experimental conditions (stringency of hybridization and washing) is demonstrated in Fig. 2.5 for the cHa-ras1 VNTR locus probe. Under stringent conditions (Fig. 2.5a) locus-specific patterns are obtained, whereas under relaxed conditions (Fig. 2.5b) multilocus banding patterns appear.

The VNTR type of RFLP was demonstrated to be due to allelic variations in repeat copy number at the minisatellite loci detected with the core probe (Jeffreys *et al.*, 1985a; Wong *et al.*, 1986, 1987). Up to now several different core sequences have been described, with homology to minisatellites in several animal species (Jeffreys, 1985; Georges *et al.*, 1988; Mariat and Vergnaud, 1992) and several plant species (Dallas, 1988; Rogstad *et al.*, 1988). This includes core probes consisting of short (14–20 nucleotides) tandem repeat motifs of random sequence composition (Vergnaud, 1989; Vergnaud *et al.*, 1991a and 1991b; Mariat and Vegnaud, 1992). Hybridization patterns obtained with different micro- and minisatellite core probes are shown in Fig. 2.6 for human, chimpanzee, horse, cattle, rat, parrot, pigeon, yeast and tomato.

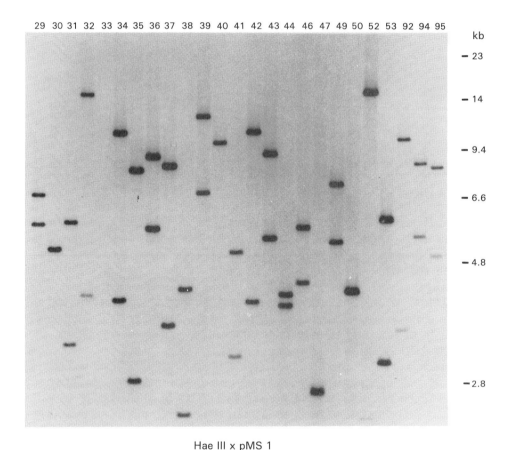

Hae III x pMS 1

Fig. 2.3. VNTR RFLP detection at the D1S7 locus (detected by probe pMS1) in Hae III digested genomic DNA from 25 different human individuals.

Sequence analysis of repeat units of particular minisatellite loci has revealed variation, which occasionally can ablate or create restriction enzyme recognition sites (Wong *et al.*, 1986; Waye and Fourney, 1990; Uitterlinden and Vijg, 1991a). This is shown in Fig. 2.4 for the D7S22 locus after hybridization analysis with pλG3. After digestion with the Hinf I restriction enzyme the expected two bands per individual show up, whereas with Hae III the father shows three bands. This third band must therefore be the result of the presence of an Hae III recognition site in the large Hinf I allele. As a result not all bands detected by a core probe will be derived from separate loci. The repeat unit sequence variation is the basis of another (very rich) source of polymorphism, characteristic for minisatellites and detectable by Southern blot hybridization (Jeffreys *et al.*, 1990) or by means of PCR analysis (Jeffreys *et al.*, 1991b).

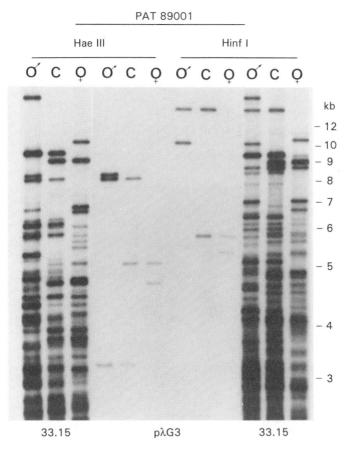

Fig. 2.4. Southern blot hybridization analysis of a human paternity case using the pλG3 probe which detects the D7S22 locus. Genomic DNAs were digested with Hae III and on the right with Hinf I. Also shown is the hybridization pattern obtained by rehybridizing the same blot with minisatellite core probe 33.15 which detects, among alleles of many other loci, the D7S22 alleles.

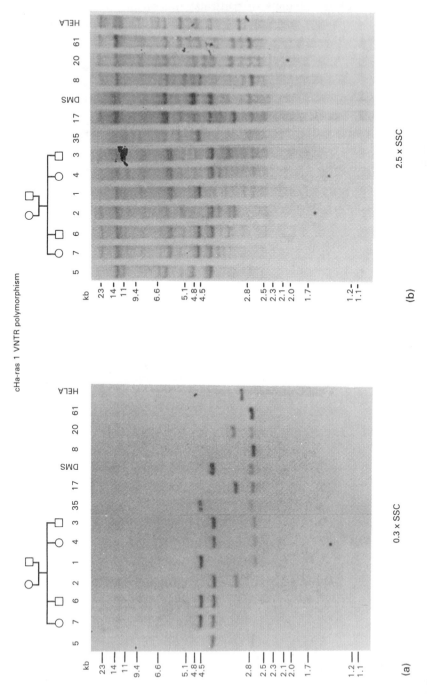

Fig. 2.5. Southern blot hybridization analysis of 5µg Pvu II digested genomic DNA from unrelated human individuals and a pedigree using a locus-specific VNTR probe (pHO6T; Capon et al., 1983) for the cHa-ras1 VNTR. The hybridization patterns shown were obtained after (a) stringent (0.3 × SSC) and (b) relaxed (2.5 × SSC); washing conditions. at 65°C of the same hybridized filter.

2.2.4.2 Microsatellites

Different from minisatellites, but a structurally very related type of tandemly repetitive sequence, are the microsatellites, the repeat unit of which has arbitrarily been set at 5 bp or smaller. Because of large variations in repeat copy number, similar to those observed for minisatellites, 'DNA-fingerprint' patterns can be obtained when microsatellite 'core-probes' are used in Southern blot hybridization analysis (Tautz and Renz, 1984; Ali *et al.*, 1986). One of the best characterized microsatellite core probes is $(CAC)_n$ (Schäfer *et al.*, 1988a and 1988b; Zischler *et al.*, 1989; Nürnberg *et al.*, 1989). In Fig. 2.6 examples are shown of the hybridization patterns obtained with probes $(AGC)_n$, $(CAC)_n$ and $(TCC)_n$ used on different species (a so-called zoo blot). Particular microsatellite loci consisting of $(GT)_n$ dinucleotides (Weber and May, 1989; Litt and Luty, 1989; Tautz, 1989; Smeets *et al.*, 1989) or different tri- and tetranucleotide motifs (Boylan *et al.*, 1990; Edwards *et al.*, 1991, 1992) can be analysed by using fully denaturing PAA gels in combination with PCR amplification techniques. They display a similar VNTR type of polymorphism to that of minisatellites, albeit with alleles of smaller length. The gel type used allows differences as small as one repeat copy of 1–3 bp to be detected. In a similar way the polydeoxyadenylate (polyA) tract of several Alu repetitive elements has been shown to be of variable size owing to a VNTR type of polymorphism of AT-rich microsatellite motifs (Economou *et al.*, 1990; Zuliani and Hobbs, 1990; Epstein *et al.*, 1990).

2.2.4.3 Large satellites

In addition to the shorter tandemly repetitive sequences much longer repetitive arrays have also been found to display polymorphism very similar to the VNTR type. These include tandemly repetitive alphoid centromeric repeats which are used to create centromere-based maps of a chromosome (Willard *et al.*, 1986; Devilee *et al.*, 1988; Wang-Jabs *et al.*, 1989; Haaf and Willard, 1992) and so-called midisatellites (Nakamura *et al.*, 1987b; Page *et al.*, 1987).

2.2.4.4 Origin of polymorphism

The majority of minisatellites described to date are composed of GC-rich repeat units which show homology to the chi sequence of *Escherichia coli*. This has led to speculation that the 'core' sequence may serve as a recombination signal to promote unequal crossing over at the minisatellite sequence (Jeffreys *et al.*, 1985a; Steinmetz *et al.*, 1986). However, several VNTRs have now been identified which are composed of AT-rich motifs (Stoker *et al.*, 1985; Knott *et al.*, 1986) including AT-rich microsatellite motifs at the 3' ends of Alu repeats (see below). In addition, many core probes detecting DNA fingerprints have been described which consist of short tandem repeat motifs of random sequence composition (Vergnaud, 1989; Vergnaud *et al.*, 1991a). This indicates that the VNTR type of polymorphism is not uniquely associated with a GC-rich chi-like motif.

Analysis of human pedigrees has shown that micro- and minisatellite VNTR loci can vary widely with respect to the frequency of generation of new alleles

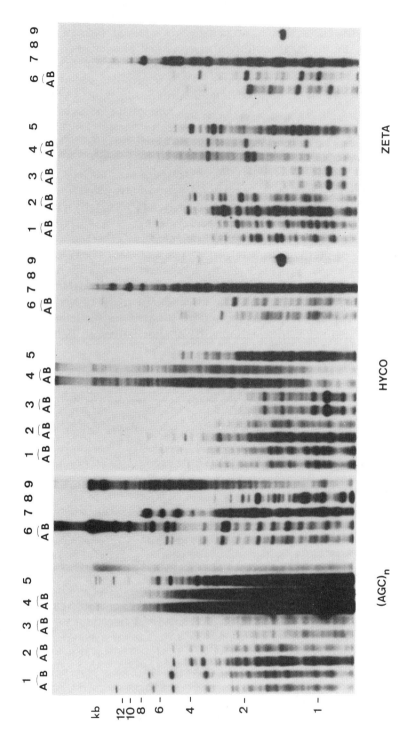

Fig. 2.6. Southern blot hybridization analysis of Hae III digested genomic DNA from different species, using different micro- and minisatellite core probes. (a) Microsatellite core probe (AGC)ₙ and minisatellite core probes HYCO and ZETA. (b) Microsatellite core probes (CAC)ₙ and (TCC)ₙ, and minisatellite core probes 33.6 and INS. The patterns were obtained by rehybridizing the same filter. A and B denote two unrelated individuals from the same species: 1, human; 2, chimpanzee; 3, horse; 4, cattle; 5, rat; 6, parrot; 7, pigeon; 8, yeast; 9, tomato.

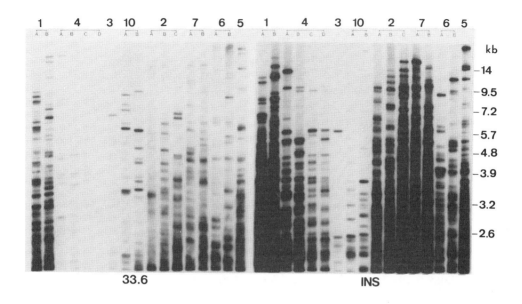

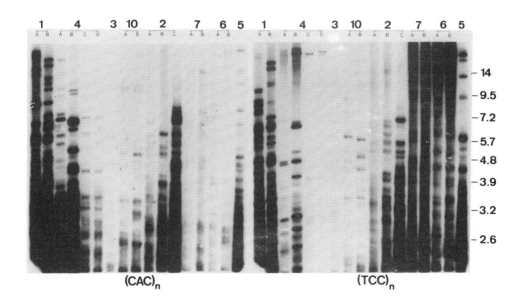

Fig. 2.6. (*continue*d).

due to mutation (Jeffreys et al, 1988, 1991; Nürnberg *et al.,* 1989). To determine the origin of the mutation, pedigrees in which mutant minisatellite alleles are observed were analysed with markers flanking the VNTR locus undergoing mutation. This showed that no homologous chromosomal areas were exchanged and thus that unequal crossing-over recombination events are not a common cause for this type of sequence variation (Wolff *et al.,* 1989). Instead, polymerase slippage replication errors are a more likely source of the variation observed. Further evidence for this was obtained by the creation of length variants of microsatellite stretches by *in vitro* synthesis (Schlötterer and Tautz, 1992).

2.2.4.5 Evolutionary background
Sequences, homologous and probably structurally related to the micro- and minisatellite sequences, have been found in a wide range of taxonomic groups (Tautz and Renz, 1984; Kashi *et al.,* 1990; Uitterlinden *et al.,* 1989b; Turner *et al.,* 1990; Schlötterer *et al.,* 1991). This is illustrated in Fig. 2.6 showing micro- and minisatellite core probes to detect homologous sequences in a variety of species. As well as in higher eukaryotes, their presence has been demonstrated in lower eukaryotic organisms such as *Caenorhabditis elegans* (Uitterlinden *et al.,* 1989b), *Musca domestica* and other insects (Blanchetot, 1991) and even unicellular organisms such as *Plasmodium* (Rogstad *et al.,* 1989; van Belkum *et al.,* 1992). It is unknown whether the sequences detected over the wide range of taxonomic groups are related with respect to function and origin. It has been speculated that the micro-, mini- and larger satellites play a role in the chromatin folding network (Vogt, 1990).

The widespread occurrence of VNTRs provides an abundant source of DNA markers also for comparative genome analysis. This was demonstrated for a number of human VNTR loci shown to cross-hybridize with mouse minisatellite loci (Julier *et al.,* 1990), for several VNTR loci isolated from bovine DNA (Georges *et al.,* 1991) and for several cetacean (whale) microsatellite loci (Schlötterer *et al.,* 1991).

2.2.4.6 Coding sequences
In a search among entries of the DNA sequence databases micro- and minisatellites have been found to be part of coding sequences (Tautz *et al.,* 1986). In particular instances the VNTR type of sequence variation can even be functional as is the case in *Plasmodium.* Here, tandemly repetitive sequence motifs encode amino acid stretches in immunodominant epitopes of membrane proteins (van Belkum *et al.,* 1992). In the human genome several examples have been found of coding sequences containing micro- and minisatellite-like repeats and displaying VNTR polymorphism. These include a tumour-associated epithelial mucin coding gene (PUM) containing a 60 bp VNTR (Swallow *et al.,* 1987), and the involucrin gene containing a 30 bp VNTR (Simon *et al.,* 1991). In exon 1 of a sarcoplasmic reticulum protein gene called HRC, $(GAG)_n$ and $(GAT)_n$ VNTRs were found (Hofmann *et al.,* 1991).

Recently discovered and clinically relevant examples of microsatellite polymorphism within the coding regions of genes involve the expansion of particular trinucleotide motifs present in human disease genes. They represent a novel mechanism of genetic disease and involve, among others, the genes for X-linked spinal and bulbar muscular atrophy (SBMA or Kennedy's disease; la Spada *et al.*, 1991), fragile X syndrome (Verkerk *et al.*, 1991; Fu *et al.*, 1991; Yu *et al.*, 1991) and myotonic muscular dystrophy (DM; Fu *et al.*, 1992; Mahadevan *et al.*, 1992). All three $(CNG)_n$ microsatellite arrays are found within the mature transcript of the gene: $(CAG)_n$ in the first exon of the androgen receptor gene in Kennedy's disease, $(CGG)_n$ in the 5' untranslated part of the FMR-1 gene in fragile X syndrome, and $(CTG)_n$ in the 3' untranslated region of the myotonin gene in DM. Whereas these microsatellite motifs display normal polymorphic variation in the human population (6–54 repeats in fragile X, 5–30 repeats in DM, and 17–26 repeats in SBMA), a substantial expansion of the array was noted in patients with the disease (200–1000 repeats in fragile X, up to 2000 repeats in DM and 40–52 repeats in SBMA). For fragile X and DM the expansion was shown to be correlated with the severity of the disease indicating a close functional relationship between the repeat region and gene regulation sites. In addition, the expansion was found to be somatically heterogeneous, probably as a result of mitotic instability, i.e. different cells show a different increment in length of the repeat array. This phenomenon has also been noted in telomeric regions containing the $(TTAGGG)_n$ repeat arrays (for a review, see Kipling and Cooke, 1992). Since the discovery of these disease genes, several other genes containing (polymorphic) trinucleotide repeat stretches have been detected in human cDNA libraries and sequence databases (Riggins *et al.*, 1992). This indicates the presence of other human genes which may undergo this new mechanism of mutagenesis giving rise to genetic disease. Indeed, most recently, Huntington's disease turned out to be the fourth disease to involve the expansion of a repeat unit, i.e. the CAG repeat (The Huntington's Disease Collaborative Research Group, 1993).

2.2.4.7 Polymorphic fraction

Since thus far only a fraction of the total number of short tandem repeat loci has been analysed for the VNTR type of polymorphism, it is unknown how many of these sequences are polymorphic in higher animal and plant genomes. Estimates of the polymorphic fraction of both micro- and minisatellite loci vary from 25% to 75% (Jeffreys *et al.*, 1985a; Nakamura *et al.*, 1987a and 1988a; Uitterlinden *et al.*, 1989a; Tautz, 1989; Armour *et al.*, 1990; Weber, 1991; Vergnaud *et al.*, 1991b)). This, in combination with their widespread occurrence in telomeric (Royle *et al.*, 1988), centromeric (Willard *et al.*, 1986) and interstitial (Luty *et al.*, 1990) regions on the chromosomes, makes these repeat sequences good candidates as markers for total genome scanning of eukaryotic genomes. Especially, coding sequences containing polymorphic micro- and minisatellite motifs present an ideal source of genetic markers, e.g. for performing genetic association studies.

2.2.5   Other interspersed repetitive sequences

Almost all genomes studied so far but especially the eukaryotic genomes have been found to contain a considerable proportion of DNA sequences which are repetitive, i.e. are present in more than two but usually in at least hundreds of copies. The proportion of the genome consisting of repeats can vary widely in different species. A particular class mainly found in lower organisms is formed by the transposons. These sequences can undergo rapid change in genome position and have been found in organisms such as bacteria, *Saccharomyces, Drosophila,* and *Caenorhabditis* and also in plant species such as *Zea mais* (McClintock, 1984).

The human genome consists of a variety of repetitive sequences that may constitute up to 40% of the complete sequence (Britten and Kohne, 1968). Because the function, if any, of most repeat sequences is unknown, current classification systems are based on empirical characteristics, such as copy number (up to 1 000 000 copies), repeat unit length (up to 6500 bp), distribution through the genome (clustered or dispersed), and local organization (tandem arrays or interspersed). In addition, multigene systems such as the MHC system and gene families such as the globin genes and homeobox genes can be considered as low copy repetitive sequences. Finally, particular sequence motifs can be found throughout the genome. Such motifs can be defined on the basis of function (e.g. zinc-finger motifs) or structure (Z-DNA stretches, palindromic recognition sites for restriction enzymes, etc.).

2.2.5.1 Polymorphism

Repetitive sequences with an interspersed organization have been shown to exhibit polymorphism in the human and animal genome (Table 2.1). In the human genome such polymorphic sites can involve Kpn I repeats (Katzir *et al.,* 1985; Economou-Pachnis *et al.,* 1985; Lakshmikumaran *et al.,* 1985; Burton *et al.,* 1985; Shyman and Weaver, 1985; Musich and Dykes, 1986; Kazazian *et al.,* 1988; Dombroski *et al.,* 1989; Woods-Samuels *et al.,* 1989), Alu I repeats (Schuler *et al.,* 1983; Lin *et al.,* 1988; Economou-Pachnis and Tsichlis, 1985; Hobbs *et al.,* 1985; Lehrman *et al.,* 1987; Vidaud *et al.,* 1989; Wallace *et al.,* 1991), and in the mouse genome members of the Sau3A and M2 family (Kominami *et al.,* 1983a and 1983b) and the B1 family (Zechner *et al.,* 1991). These polymorphisms can be the result of the absence or presence of one or more copies of a repeat unit. This will create or destroy a restriction enzyme recognition site (RSP: restriction site polymorphism) or alter the length of a given restriction fragment (insertion/deletion or variable number of tandem repeat (VNTR) polymorphism). RFLPs of interspersed repeats (in which case a single repeat unit is either present or absent at a particular locus) are not essentially different from 'classical' RFLPs; only two alleles can be discerned which have varying frequencies in a population. In general this type of polymorphism is very infrequent.

Another class of repeats found in mammalian genomes involve viral-like elements, including the so-called intracisternal A-particles (IAPs) in mice and retroviral-like elements in humans. IAPs are a closely related set of endogenous proviral-like sequences present in the genome of all inbred mice (Kuff and Lueders, 1988). They have been shown to be polymorphic between strains of mice and probes for IAP sequences have been successfully used to perform genetic linkage studies (Mietz and Kuff, 1992) and to detect a disease-causing duplication in the mouse (Brilliant *et al.*, 1991). In humans different classes of retroviral elements, occurring in copy numbers of 1000–2000, have been found including the RTVL-H (also known as HERV-H) and the HERV-A elements. The low copy number of these repetitive elements allows them to be used for characterization of somatic cell hybrids (Sugino *et al.*, 1992) and isolation of region-specific cosmid clones (Meulenbelt *et al.*, 1993). Infrequent polymorphisms due to homologous recombination of LTRs have been observed for the RTVL-H elements (Mager and Goodchild, 1989).

Sequence analysis of individual members of many of these repeat families has revealed basepair substitutions within repeat units, suggesting a widespread occurrence of polymorphic basepairs within repeat loci in the human population. Since point mutations will only rarely alter a specific restriction enzyme recognition site, this kind of polymorphism is difficult to detect by Southern blot analysis, especially in view of the sometimes extremely high copy number of these repeat families.

2.2.5.2 PCR analysis

Single 'consensus' primers for repetitive sequences in a PCR reaction can be used for the simultaneous analysis of many interrepeat regions (see Fig. 2.7). One such primer is capable of detecting many loci simultaneously and therefore of great value in parallel genome analysis. Repetitive sequences flank the region of analysis and hence this type of PCR analysis is referred to as interrepeat PCR (irPCR). The technique was first established for the analysis of human–hamster somatic cell hybrids containing single human chromosomes based on known repetitive sequences in the human genome such as Alu I (Nelson *et al.*, 1989) and Kpn I or LINE repeats (Ledbetter *et al.*, 1990a). The primers can be chosen such that only the human Alu I repeats are detected but not the homologous but more distantly related rodent B1 and/or B2 repeats. Similar irPCR primers have been developed for the analysis of somatic cell hybrids of mouse chromosomes on a hamster (Cox and Lehrach, 1991; Cox *et al.*, 1991; Simmler *et al.*, 1991) or human background (Herman *et al.*, 1991). Besides primers specific for known dispersed repetitive elements also primers homologous to microsatellite sequences such as $(GA[C/T]A)_n$ have been successfully applied (Cox *et al.*, 1991). As a result of the dispersed distribution of the repeats, the collection of fragments obtained from a hybrid containing one human chromosome or part thereof constitutes a specific 'fingerprint' or karyotype. By comparing such patterns small variations in chromosome constitution can be analysed by comparison of electrophoretic separation patterns of the PCR products (Ledbetter *et al.*, 1990b) or by *in situ* hybridization (Lichter *et al.*, 1990). Such panels of chromosome-specific irPCR fragments have been established for several human chromosomes including chromosome 6 (Meese *et al.*, 1992) and chromosome 10 (Brooks-Wilson *et al.*, 1990).

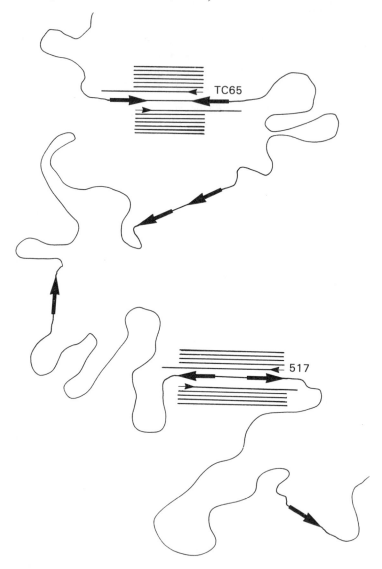

Fig. 2.7. Schematic representation of interrepeat PCR using primers TC65 and 517 (small arrows) specific for the human Alu I repeat (large arrows).

Two different types of polymorphisms have been detected in the analysis of Alu I repeats. One consists of a microsatellite type of VNTR polymorphism at the 3′ end of the Alu repeat unit consisting mainly of $(T[T/A]A)_n$ but also including other types of repeat units (Economou *et al.*, 1990; Zuliani and Hobbs,

1990; Epstein *et al.*, 1990). The other type is based on the presence or absence of a priming site for the consensus primer as a result of either basepair polymorphism or presence/absence of a complete repeat unit (Sinnett *et al.*, 1990; Zietkiewicz *et al.*, 1992). Similar types of polymorphism might be found for the LINE type of repeat since these have also been found to contain polyA stretches at the 3′ end (Economou *et al.*, 1990).

Besides using primers specific for known repetitive elements, 'random' primers have also been applied in irPCR (Williams *et al.*, 1990; Welsh and McClelland, 1990) including primers as short as 5 bp (Caetano-Anollés *et al.*, 1991). This approach has been used successfully in several species to detect polymorphisms based on the presence or absence of amplified fragments. Such polymorphisms can be used for strain identification and genetic mapping, for example, in the mouse (Welsh *et al.*, 1991). Since no locus information is available such markers are referred to as random amplified polymorphic DNA (RAPD). In particular segregating populations (i.e. populations of sibs from a single cross) the method has been used to develop markers in a region-specific manner (Michelmore *et al.*, 1991; Giovannoni *et al.*, 1991). Disadvantages of the RAPD system are the limited number of fragments, the low fraction of variant bands detected per primer and the low informativity of the dominant bi-allelic polymorphism. These characteristics require a very large number (several hundreds to several thousands) of primers and PCR reactions per individual for comprehensive genome analysis. Furthermore, their lack of species specificity effectively constrains their application in chromosome fingerprinting/karyotyping. Although Mendelian inheritance has been observed, an excess of mutant bands was found in paternity analysis of primates (Riedy *et al.*, 1992). Further evaluation by cloning and sequence analysis of the polymorphic loci detected by RAPD PCR will allow the origin and location of these sequences to be determined and the usefulness of this approach for genome analysis to be evaluated.

In summary, interrepeat PCR based on known repetitive elements seems a powerful tool in parallel genome scanning approaches. A single primer can detect hundreds if not thousands of potential marker loci and the method can be applied in a chromosome-specific manner. Combination of several primers specific for different types of repetitive sequences will allow comprehensive genome analysis and thereby total genome comparison among individuals. This will be discussed further in Chapter 5.

2.2.6 Total genome analysis with DNA markers

From the above it can be concluded that repetitive sequences are a valuable source of polymorphic markers. Such markers can be employed in a locus-specific manner by serial analysis for their local repeat copy number. Alternatively, repeat loci can be used as anchor points in total genome scanning of individuals by using probes that detect many repetitive sequences simultaneously, e.g. micro- and minisatellite core probes, interrepeat amplification primers.

The essence of genome comparison studies, e.g. linkage mapping of disease genes or the analysis of tumour genomes, is to scan the genome for DNA sequence variation at evenly spaced points along each of the chromosomes. DNA probes, detecting such points of variation, can identify chromosomal regions containing particular genes by following the transmission of particular variants in pedigrees, or in studies of tumourigenesis, ageing and mutagenesis to identify genome alterations. The efficacy of such an analysis depends in part on the density of markers which constitute the genome map.

A major application of the results to be obtained in the genome mapping and sequencing endeavour will be to compare genome sequence organization in different strains or individuals to identify and isolate specific genes. The approach to assess individuality by sequencing complete individual genomes, however, is unfeasible and cost ineffective in view of the high level of similarity and the resulting redundancy of the data generated. A more efficient approach is to focus directly on the sequence differences, which can then be used as marker loci in linkage and/or association studies aimed at identifying genes responsible for medical conditions and other traits in humans, animals and plants.

When more complex problems are to be addressed, such as analysing genetic heterogeneity or polygenetic inheritance, the identification of the DNA sequence variations involves substantial efforts. Such studies require the complete genome of interest to be densely populated with (highly) polymorphic markers spread over the genome and statistical methods that enable exploitation of such a rich source of genetic information (Lander, 1988; Lander and Botstein, 1986, 1989). The feasibility of such an approach has been demonstrated by some preliminary studies, resolving particular quantitative traits, influencing for example fruit mass in tomato (Paterson *et al.*, 1988) and hypertension in rats (Hilbert *et al.*, 1991), into Mendelian factors and to map these quantitative trait loci (QTL) on the chromosomes.

In view of the complexity of the higher eukaryote genome it would be advantageous to be able to monitor sequence variation in parallel on a total genome scale, i.e. at multiple sites simultaneously. Even when the number of fragments per individual DNA sample can be increased to several hundreds to a thousand by using two-dimensional electrophoretic separation of fragments, this will only allow small genomes (up to 5 Mbp) to be analysed comprehensively, i.e. to analyse all restriction fragments. For larger genomes the number of restriction fragments generated will be too high. Therefore techniques have to be employed which visualize selected sets of polymorphic DNA sequences dispersed over the genome of interest. Such sequences can include families of repetitive sequences or can consist of recognition sites for restriction enzymes. Techniques to visualize either of these can be divided in three categories.

(1)   Hybridization analysis of electrophoretically separated genomic DNA fragments using repetitive sequences as probes (Southern, 1975; Gusella *et al.*, 1980 and 1982; Jeffreys *et al.*, 1985a; Porteous *et al.*, 1986; Uitterlinden *et al.*, 1989a; Sugino *et al.*, 1992; Meulenbelt *et al.*, 1993).

(2)　Interrepeat PCR using primers specific for particular repetitive motifs either derived from known repetitive elements (Nelson *et al.*, 1989; Ledbetter *et al.*, 1990) or chosen randomly (Williams *et al.*, 1990).

(3)　Selective end labelling of restriction fragments for example by first applying enzymes with relatively few recognition sites (rare cutters) followed by labelling and digestion with a more frequently cutting enzyme. This was first described for analysing cosmids (Carrano *et al.*, 1989) but has also been applied for the analysis of total mammalian genomes (Yi *et al.*, 1990; Hatada *et al.*, 1991).

These methods can be combined to analyse comprehensively any given genome of interest. Since the occurrence of repetitive sequences varies with different genomes, both the hybridization approach and the irPCR approach are somewhat limited in their application. This is, however, counterbalanced by the amount of polymorphism they can detect in contrast to the limited DNA sequence polymorphism in restriction enzyme recognition sites. These factors, however, remain to be evaluated in the practice of analysing genomes.

2.3　COMBINING GENETIC AND PHYSICAL MAPPING

For several organisms with relatively small genomes such as *Escherichia coli, Saccharomyces cerevisiae, Caenorhabditis elegans* and *Drosophila melanogaster* the first genetic and physical maps have already been created and the sequencing of large contiguous stretches of DNA sequence is well underway. For *C. elegans,* for example, the genome is estimated to span approximately 100 Mbp or 1×10^8 bp, 95 Mbp of which is now present in cloned cosmids and YACs, and 90 Mbp has been positioned along the chromosomes. The complete sequencing of only three cosmids spanning 120 kb has already resulted in the identification of new and hitherto unknown genes and has given insights into gene organization and gene density in this organism (Sulston *et al.*, 1992). Another example is the recently completed sequence of chromosome 3 of the yeast *Saccharomyces cerevisiae* by a consortium of 35 European laboratories (Oliver *et al.*, 1992). The sequence comprises 315 357 bp with 182 open-reading frames, most of which (116) appeared to represent hitherto unknown genes. Besides being useful for comparative analysis these projects will also function as examples of how to achieve the goals of the larger projects on higher animals and plants in the most efficient way.

The coordinated use of genetic and physical mapping techniques and the computerized evaluation of the results is now culminating in several genome projects. Of these, the genome projects for smaller organisms, like the ones mentioned above, are likely to be finished before the year 2000. For higher organisms the human genome project naturally attracts most attention (Watson, 1990; Cantor, 1990) while comparable genome projects for animals and plants are also underway (Roberts, 1990b; Womack, 1992). In general such projects will lead to:

(1)   a very dense genetic map with (highly) informative polymorphic markers each 1–2 Mbp,

(2)   ordered libraries of cloned fragments in YACs and cosmids and correspondence between the genetic and physical map through areas defined by genetic markers, and

(3)   a complete consensus sequence with all genes identified.

For the smaller genomes the information gathered can be managed by rela- tively simple means. However, for projects analysing genomes the size of the human genome, it is likely that numerous different laboratories will contribute data to compile the genetic and physical map. Currently, genetic mapping of the human is centred around the initiative of the CEPH consortium (Dausset *et al.,* 1990). In this initiative, a given panel of human pedigrees is being used by several different research groups around the world for typing all available genetic markers with a known physical map location. All the genetic typing data are collected in the CEPH database which will provide an increasingly more dense genetic map of each of the human chromosomes. Already, a comprehensive genetic linkage map of the human genome has evolved from this world wide collaboration based mainly on the use of RFLPs and some $(GT)_n$ repeats (NIH/CEPH Collaborative Mapping Group, 1992; Weissenbach *et al.,* 1992).

Apart from genetic markers being placed on the physical map, a common lan- guage for physical mapping of the human genome in different laboratories has also been proposed (Olson *et al.,* 1989). The concept is based on the selection of sequences of 200–500 bp which can be uniquely defined by two flanking PCR primers. Such 'sequence tagged sites' (STSs) will allow rapid integration of physical mapping data collected in different laboratories. A logical extension of STSs would be to have them include polymorphic sequences, so that they can be used both for genetic mapping and for identifying particular areas along the chromosomes.

A prerequisite for comprehensive genome analysis is the ability to interrelate data from the physical and the genetic maps. In addition to the use of STSs, this can be accomplished by the use of identical types of sequences in physical mapping and in genetic analysis such as cloned and mapped DNA markers. An alternative system for genome mapping, the reference library system, has been proposed as a more efficient way to collect and distribute sequence information on genomes (Lehrach *et al.,* 1990). In this system, an ordered array (grid) of chromosome-specific cosmid clones and an array of YAC clones covering the complete human genome are provided on a hybridization membrane. Identical copies of the membrane are sent to different users who each hybridize the grid with their probe of interest and report back the coordinates of the hybridizing cosmid/YAC clones. This will eventually result in the localization of all genetic markers to particular cosmids and YACs. In parallel with this, the overlap of the cosmids themselves is determined by hybridizing the membrane with a set of short oligonucleotides 12–15 nucleotides long which hybridize at multiple sites in the genome. This will result in only one cosmid out of several showing hybridization with a particular oligonucleotides. Thus, each cosmid can be given a unique signature of hybridization with a particular set of these oligonucleotides.

The principle of the oligonucleotide hybridization signature can be extended by taking shorter oligonucleotides (6–8-mers) to a level of sensitivity that allows the sequence of the fragment to be determined (see also Chapter 1, section 1.2.1). For any of these applications the number of oligonucleotides necessary to perform mapping and sequencing is determined by the size (and sequence) of the genome of interest. The combination of reference libraries and oligonucleotide hybridization analysis thus provides a high-resolution, sequence-based mapping system and in the longer run even a sequencing system (Drmanac *et al.,* 1989, 1993).

A major advantage of this mapping and sequencing system is the optimal data output due to the parallel nature of the analysis. Such parallel analysis is accomplished by using sequences which recognize multiple targets in the sequence of interest. For the reference library system short random oligonucleotides are used. However, also defined repetitive sequences showing suitable genome frequencies have been used in this physical mapping approach (Lehrach *et al.,* 1990). Additional examples of the use of repeats in determining overlap of clones are the use of $(GT)_n$ repeats and sets of repetitive sequences (by hybridization with total genomic DNA) in the mapping of YACs (Bellanné-Chantelot *et al.,* 1992) and chromosomes (Stallings *et al.,* 1990). Since repetitive sequences in general and the micro- and minisatellites in particular have been shown to display extensive polymorphism they can also be used for genetic analysis. Therefore, they are good candidates to form a link between physical and genetic mapping in the parallel mapping approach described above combined with two-dimensional DNA typing using these repeats as probes.

2.4 STRATEGIES FOR IDENTIFYING GENETIC TRAITS

DNA sequence variation can be studied on a population level to reveal the basis of phenotypic variations observed in different individuals. The most striking variations in this respect are disease-causing gene variants. Attempts to identify such gene variants without *a priori* knowledge of the biochemical basis of the defect, and therefore no clue as to what protein might be involved, encompass little less than looking for a needle in a haystack without knowing what the needle looks like.

The discovery of DNA sequence variants as genetic markers (see section 2.2.3), more or less randomly dispersed over the genome, gave a new impetus to what is known as genetic linkage analysis. In linkage analysis a set of families is typed for a series of genetic markers. Co-segregation of a marker with a particular trait, such as a disease, more often than by random chance indicates linkage. According to Mendel, variants of two loci have a 50%

chance of being inherited together. If the two loci are in close proximity on the same chromosome this frequency increases. The probability for each marker to be linked to the genetic disease occurring in the family under study can be calculated. The result is called the 'lod score' and is usually given at a range of recombination values. The lod score is the logarithm to base 10 of the ratio between two probabilities: the probability that a given segregation pattern can be observed when two loci are linked and the probability of the observations in the absence of linkage. A lod score of 3 (1000:1) or higher is usually considered as evidence that the two loci are linked. Lod scores can easily be calculated by using computer programs (Lathrop and Lalouel, 1984; te Meerman, 1991; Ott, 1991).

The chance of recombination to occur is lower for a smaller distance between the disease locus and the marker locus. This phenomenon is exploited to measure the 'genetic distance' between the two loci: if no or very few recombinations are observed in pedigrees segregating for the disease it can be assumed that the marker locus and the disease locus are very close to each other on the chromosome. Scoring of such recombination events between a set of defined markers and the phenotype of interest is used to define the chromosomal area harbouring a particular gene of interest. From this moment onwards the actual 'positional cloning', a term coined by F. Collins (Collins, 1992) and formerly known as 'reversed genetics', should commence.

Although positional cloning by linkage analysis has been very successful in the identification of simple genetic traits, such as monogenic human diseases (e.g. Duchenne muscular dystrophy, cystic fibrosis, neurofibromatosis type I), it is not necessarily the method of choice for analysing all diseases with a genetic component. For example, complex human diseases such as Alzheimer's disease, schizophrenia and manic depression proved to be difficult to analyse at first by linkage strategies. Indeed, problems can arise in case of heterogeneity, i.e. when different mutant alleles from either the same or different loci give rise to the same phenotype, or when more than one gene contributes to the same character, i.e. quantitative traits such as height in humans and milk production in cattle. Although both heterogeneous and quantitative traits are in principle amenable to the lod score method (Lander and Botstein, 1989), complicating factors such as an unclear mode of inheritance, due for example to difficulties in accurate diagnosis, were responsible for some apparent failures (Marx, 1990).

Together, the quality of the pedigree material, the availability of higher-density genetic maps and a clear insight into the mode of inheritance determine the success rate of the linkage approach in identifying disease gene loci. Certain advances in one of these factors can compensate for the others, but not always. In general, when one is looking for genes that may contribute not more than 10% to a given phenotype, they may be difficult to identify by linkage analysis. In such cases it may help to increase the density of the genetic map, i.e. using many more markers spread over the genome. In view of the recently obtained high-density maps of the human genome (see above), a fresh start has been proposed for such diseases as manic depression and schizophrenia (Gershon *et al.,*

1990). In certain instances, however, the linkage approach may simply not be the method of choice for solving complex diseases.

Besides linkage analysis other strategies can be applied including association analysis. Whereas in linkage analysis one tries to find unknown genes by analysing families using anonymous markers as a tag to mark the chromosomal region that carries the trait, association analysis singles out a particular gene or family of genes for which there are biological reasons to think they might be involved in the particular phenotype that belongs to the genetic trait studied. Since association studies focus on populations rather than on families the use of polymorphic markers is much more restricted. Indeed, whereas family members are closely related, population members are much more genetically diverse with, consequently, a much higher possibility of a recombination event breaking the link between marker and trait of interest. Only when the polymorphism is in or very near the gene of interest, or in linkage disequilibrium with the gene, is the marker useful (see also Chapter 4).

The best example of association studies is the observed correlation between particular MHC (major histocompatibility complex) alleles and human diseases (e.g. diabetes, arthritis, coeliac disease) (for a review, see Thomson, 1988). The genes of the MHC (called HLA in humans) are clustered in a region of over 3000 kb on the short arm of chromosome 6. MHC molecules play a key role in the generation and regulation of immune responses. Since in most of the diseases for which association with MHC alleles has been found an immune defect is probably involved, there is some inherent logic in the observation that particular MHC haplotypes influence susceptibility. However, it should be noted that as yet there is no information as to the actual gene(s) involved, that is, whether the MHC loci are themselves the susceptibility loci or are linked to them. To some extent this is due to the high degree of sequence variations among MHC alleles and the impossibility of comprehensively scanning the region for any given sequence variation. Therefore it seems important to scan accurately the entire region for DNA sequence variation associated with particular disease phenotypes.

A special form of association analysis is the search for loss of heterozygosity (LOH) in various human cancers. By now there is overwhelming evidence that cancer is a genetic disease and finds its causes in changes in the DNA sequence organization. However, unlike most other genetic diseases cancer expresses its genetic characteristics also at the somatic level where the affected clonally derived tissue can be genetically distinguished from the unaffected. The best example of genes that confer susceptibility to cancer are the so-called tumor suppressor genes (Cavenee et al., 1989; Sager, 1989). It is now generally believed that individuals with an inherited recessive mutation at such loci are predisposed to cancer; a tumour would occur once the dominant wild-type allele is eliminated (loss of heterozygosity). Interestingly, such disease-predisposing loci can be found by comparing DNA from the tumour tissue with that of normal unaffected tissue from the host. In this case one simply tries to associate (the loss of) DNA sequence variants with tumour phenotypes, irrespective of a family relationship between the various patients with that particular cancer. Needless to say, such a strategy relies entirely on the availability of rapid genome scanning techniques.

2.5   THE ROLE OF TWO-DIMENSIONAL DNA TYPING IN GENOME ANALYSIS

The analysis of a complete genome for sequence variation can be done by serial, locus-specific analysis but this is much less efficient than the parallel, multilocus type of analysis. For the latter several formats have become available in the past few years (among which 2D DNA typing) mainly based on a combination of electrophoretic separation of DA fragments and hybridization or PCR analysis. Examples of such multilocus parallel processing methods for genome analysis ('genome scanning') include hybridization analysis of mouse DNA using an intracisternal A particle-specific probe (Brilliant *et al.,* 1991; Mietz and Kuff, 1992) and inter-repeat PCR analysis of human DNA (Zietkiewicz *et al.,* 1992). An alternative form of genome scanning is based on hybridization reactions involving complete genomes. Rather than identifying variation by analysing genomes at multiple regularly spaced sites, these methods allow the purification of regions in common or different between two complex genomes. Recently two methods have emerged as most promising in this respect: genomic mismatch scanning (GMS; Nelson *et al.,* 1993), which is designed to identify DNA sequences two genomes have in common, and representational difference analysis (RDA; Lisitsyn *et al.,* 1993) which seeks to identify the differences between two genomes.

The GMS method is an extension of earlier work on modifying the phenol emulsion reassociation technique (Sanda and Ford, 1986; Casna *et al.,* 1986). In this method each of the genomes to be compared is first digested with a restriction enzyme producing fragments with protruding 3 ends (for protection against exonuclease III at a later stage). Then, one of the two genomes is modified by methylation (at 5-GATC-3 sites). The two genomes are then hybridized to one another, which will result in two types of double-stranded fragments: fragments with two strands derived from the same (homoduplex fragments) or fragments from different genomes (heteroduplex fragments). The homoduplex fragments are either both methylated or completely unmethylated and are digested with Dpn I (cutting only fully methylated GATC sites) and Mbo I (cutting only unmethylated GATC sites), respectively, to produce 5 protruding ends. Heteroduplex fragments containing mismatches (due to sequence differences between the two genomes) are eliminated by enzymatic treatment with mismatch-recognizing enzymes (mut-H, -L and -S). These enzymes recognize only hemimethylated fragments and subsequently introduce a single-strand nick in the unmethylated strand at 5-GATC-3 sites. Finally, the mixture is treated with ExoIII, an exonuclease which recognizes nicks, blunt ends and 5 protruding ends (but not 3 protruding ends). Hence, only perfectly matched fragments with 3 protruding ends are recovered. The net result is the purification of mismatch-free fragments consisting of one strand of each genome which are subsequently used as a probe on an ordered genomic library representing a physical map of the organism. This will identify all regions of identical sequence between the two genomes analysed. The method has been successfully applied in the analysis of the yeast genome to isolate such fragments but analysis of more complex, repeat-rich genomes, such as the human genome, awaits further technical improvements.

In contrast the RDA method is based on 'subtractive hybridization', a method already known, but usually yielding only 10100-fold enrichment of fragments different between genomes. However, with RDA enrichments of 10 can be achieved. To accomplish this, first the complexity of the hybridization reaction is lowered by preparing representative fractions of the genomes to be compared. Such fractions (less than 10% of the genome) are obtained from restriction fragments (ligated to oligonucleotide adapters) by PCR amplification, thereby selecting for only the small (kb) restriction fragments. Fragments derived from the genome lacking the target (driver) are then hybridized in great excess to fragments from the genome with the target (tester). The latter fragments were provided with new adapters. Reannealed tester fragments (supposedly not present in the driver genome) are then subsequently exponentially amplified and the whole procedure can be repeated using different PCR adapters. This will lead to a purification of fragments either absent in the driver DNA, or of fragments which were not amplified during the preparation of the representation fraction. In this way deletions can be identified in the driver DNA. Alternatively, restriction site polymorphisms (RSP) between the two genomes can be found because their presence leads to a fragment too large to allow amplification in the genome lacking the recognition site.

The method has been successfully applied to identify RSP's between human DNAs and can also be used to identify specific pathogens in genomic DNA and abnormalities in tumour DNA. RDA may find a major application in the identification of previously unknown sequence differences (mainly insertions or deletions). For rapid screening of multiple polymorphic sites, however, it offers no particular advantage.

A third example of entire genome hybridization analysis is comparative genomic hybridization (CGH; Kallioniemi *et al.,* 1992). In this method the two genomes to be compared (i.e. tumour and normal DNA) are screened by fluorescence microscopy analysis of metaphase spreads for local differences in sequence copy number, i.e. amplifications and deletions. The two genomic DNAs are tagged with green and red fluorescent labels, respectively. Subsequently, they are simultaneously hybridized to a normal metaphase spread in the presence of unlabelled Cot-1 blocking DNA to suppress hybridization of repetitive sequences (e.g. at centromers). Quantification of the ratio of green to red fluorescence by digital analysis of fluorescence microscopy images, allows measurement of the relative amounts of tumour and normal DNA bound at a given chromosomal locus. Amplifications in tumour DNA will produce an elevated ratio while deletions lead to a decreased ratio. In this way a physical map is obtained of differences in DNA sequence copy number (deletions but mainly amplifications, ranging in size from a few hundred kb to several Mbp) as a function of chromosomal localization.

Table 2.2. Genome scanning methods

Method[A]	Techniques	Number of loci detected	Variance[B]	Type of polymorphism[C]	Species-specific	Locus-information
SERIAL LOCUS-SPECIFIC						
RSP probe	gel + probe hybridization	1/probe	5–10%[D]	RSP	yes[E]	yes
VNTR probe	gel + probe hybridization	1/probe	25%[D]	VNTR, RSP	yes[E]	yes
(GT)$_n$ repeat	PCR + gel	1/primer set	20%	VNTR	yes[E]	yes
PARALLEL MULTI-LOCUS						
1D electrophoresis						
DNA fingerprinting	gel + core probe hybridization	20/probe	99%	VNTR, RSP	no[F]	no[G]
LTR (IAP) probe[H]	gel + probe hybridization	20–100/probe	1–50%	INS/DEL, RSP	yes	no[G]
inter-repeat PCR[K]	PCR + gel	40/primer	50%	INS/DEL	yes[F]	no[G]
RAPD-PCR	PCR + gel	10–30/primer	20%	INS/DEL	no[F]	no[G]
2D electrophoresis						
2D DNA typing	gel + core probe hybridization	400–500/probe	80%	VNTR, RSP, SP	no[F]	no[G]
RLGS	end-label + gel	100–2000/gel	1–5%	RSP	no	no[G]
Solution hybridization						
GMS	hybridization + enzyme	total genome	[L]	INS/DEL	no	yes
RDA	hybridization + PCR	10% of genome	[L]	RSP, INS/DEL	no	yes
CGH	label + hybridization	total genome	[L]	FCN	no	yes

[A] core = micro- or minisatellite sequence; LTR = long terminal repeat; RAPD = random amplified polymorphic DNA (Williams *et al.*, 1990; Caetano-Anollés *et al.*, 1991); RLGS = restriction landmark genomic scanning (Hatada *et al.*, 1991); GMS = genomic mismatch scanning (Nelson *et al.*, 1993); RDA = representational difference analysis (Lisitsyn *et al.*, 1993); CGH = comparative genomic hybridization (Kallioniemi *et al.*, 1992).

[B] Percentage of variant fragments (in terms of presence or absence detected) among unrelated individuals.

[C] RSP = restriction site polymorphism; VNTR = variable number of tandem repeats; INS/DEL = insertion and deletions; SP = (restriction enzyme independent) sequence polymorphism; FCN = fragment copy number.

[D] As determined by screening recombinant libraries; for scanning purposes a preselection of informative markers should be made.

[E] Sets of hundreds of locus-specific probes/primers should be developed for each species.

[F] Optimal set of a few probes/primers should be empirically determined.

[G] Genetic maps can be constructed which are used to localize fragments in co-segregation analysis.

[H] Mietz and Kuff (1992); Brilliant et al. (1991).

[K] Zietkiewitcz et al. (1992).

[L] Only differences (RDA and CGH) or sequences in common (GMS) are identified.

In Table 2.2 an overview is presented of genome scanning methodologies currently available. They can be divided into two groups based on either entire genome hybridization analysis, or on gel electrophoretic separations followed by detection of particular subsets of fragments. The first group includes genome mismatch scanning (GMS; Nelson *et al.*, 1993) and representational difference analysis (RDA; Lisitsyn *et al.*, 1993) which are based on purifying either differences (RDA) or sequences in common (GMS) between two genomes. These methods therefore do not allow analysis of the genome at regularly spaced intervals for variations. Alternatively, CGH (Kallioniemi *et al.*, 1992) assays for differences in DNA sequence copy number along the chromosomes by fluorescence microscopy. The method provides direct information on the locus involved but at relatively low resolution (differences down to a few hundred kb are detected).

The group of genome scanning methods based on gel electrophoretic separation offers much higher resolution in terms of differences that can be observed but the location of possible differences is not *a priori* known. The methods are based on either a serial locus-specific analysis or on a parallel multilocus analysis. The first one includes the now classic approaches to linkage analysis using either RSP probes detecting only two alleles, VNTR probes detecting more than two alleles per locus (both type of probes analysed by Southern hybridization) and the PCR analysis of the $(GT)_n$ type of microsatellite VNTRs (Hearne *et al.*, 1992). In serial approaches hundreds of locus-specific probes have to be isolated from recombinant libraries and only a fraction of them will detect polymorphic loci which can be assayed in genome scanning studies.

The multilocus approaches include techniques based on one-dimensional or two-dimensional separation patterns, including 2D DNA typing, inter-repeat PCR analysis of dispersed repetitive sequences, or using end-labelling of particular restriction fragments. They have in common that many variant fragments are analysed but that no immediate locus-information is obtained upon identification of a fragment of interest. This book is specifically occupied with the comprehensive genetic analysis of large genomes that can be accomplished by combining a two-dimensional electrophoretic method, to resolve and detect simultaneously the maximum number of fragments, with hybridization or PCR analysis, using probes/primers to detect polymorphic dispersed repetitive sequences. The application of DGGE in 2D DNA analysis ensures comprehensive detection of polymorphisms at any site in the genome (Chapter 1).

Two-dimensional DNA typing is capable of linking detected sequence variants to physical mapping information in several ways. First, all fragments constituting a relatively small genome (or large region from a genome) can be completely resolved and simultaneously analysed for DNA sequence variation. Second, for larger genomes a fraction of sequences which have known physical map locations can be analysed simultaneously as anchor points for DNA sequence variation. Ideal anchor points in this respect are repetitive sequences which have a dispersed distribution through the genome.

Two-dimensional DNA typing can be applied in many areas of genome research. These include the following.

(1)   The direct mutational analysis of DNA fragments (generated by restriction enzyme digestion or by PCR amplification) derived from contiguous stretches of DNA up to a length of several million basepairs. Examples are a single genome in the case of lower organisms (Fischer and Lerman, 1979a) or large genomic regions such as a gene with introns and exons, in the case of higher organisms.

(2)   Hybridization and/or PCR analysis of large genomes using repetitive sequences specific for polymorphic loci in genetic studies or in studies of somatic instability. Examples are the application of core probes (Uitterlinden *et al.*, 1989a) and interrepeat PCR (Uitterlinden *et al.*, 1991a).

(3)   Analysis of complex mixtures of DNA molecules derived from populations of different individuals or different species. Examples are the analysis of 16S RNA to fingerprint populations composed of different microbial species (Muyzer *et al.*, 1993), and the analysis of cDNA populations specifying mRNA populations.

In the next chapter the methodology used to produce and interpret two-dimensional separation patterns will be provided. In Chapter 4 some applications of 2-D DNA typing will be reviewed in detail. Finally, in Chapter 5 we will discuss the future prospects of the 2D DNA typing concept.

3

General methodology

3.1 INTRODUCTION

The standard format of two-dimensional DNA typing is fully based on the system first described by Fischer and Lerman (1979a and b). The first dimension electrophoresis involves size separation either in agarose or polyacrylamide gels and the second dimension electrophoresis is performed in a polyacrylamide gel containing a gradient of the denaturants urea and formamide. The choice of the first dimension matrix is dictated by the average size of fragments to be separated. If these are larger than 1–2 kb, agarose is more suitable than polyacrylamide for obtaining optimal separation. If fragments widely varying in length have to be separated (e.g. ranging in length from 100 bp up to 3 kb), improved size separation can be achieved by using linear gradients of polyacrylamide. Usually a linear gradient of 4–9% polyacrylamide is sufficient to obtain good resolution. In addition, improved resolution over a wide range of fragment sizes can be obtained by using longer gels (i.e. twice the length of the normal gels).

For the second dimension separations 6% polyacrylamide gels are currently used. If very small fragments have to be analysed (100–500 bp) it is advisable to use higher percentages of acrylamide. However, the percentage of polyacrylamide only slightly influences the focusing of fragments. Moreover, the higher the concentration of polyacrylamide the more difficult it becomes to transfer the resulting separation patterns to hybridization membranes.

A key factor in the generation of optimal 2D DNA separation patterns is the quality of the genomic DNA. We have found that already a slight amount of single-strand breaks in the genomic DNA that has no visible effect on the first dimension (size) separation, negatively influences separation in the second dimension. Such breaks will cause spots in two-dimensional separations to be streaky and/or fuzzy.

In the following sections a detailed description is given of all the steps in the procedure of two-dimensional DNA typing, from the isolation of genomic DNA to hybridization analysis. The protocols provided can be used for obtaining optimal 2D DNA typing patterns in hybridization analysis of higher eukaryotes using micro- or minisatellite core sequences and other probes. They can also be applied, however, for analysing lower organisms either directly by ethidium bromide staining or by hybridization analysis with a variety of probes.

3.2   METHODS

3.2.1   DNA isolation
3.2.1.1   General aspects
DNA isolation protocols are provided for common sources of DNA, namely blood and sperm for humans and animals, and bulbs and leaves for plants. It is our experience that the isolation protocols described result in genomic DNA of sufficient quality for 2D separation. However, other protocols could also suffice. In general, DNA quality should be checked by alkaline agarose electrophoresis (see section 3.2.1.4). This will indicate the presence and amount of single-strand breaks which can negatively influence the quality of the 2D spot patterns.

3.2.1.2   Blood
This protocol has been used to isolate DNA from blood from humans, cattle, horses, pigs and several primate species. Note that for non-mammalian species, such as birds, fish and reptiles, the red cells are nucleated and hence contain DNA. The red cell lysis procedure should be omitted for these species.

For 10 ml fresh whole blood, collected in a tube containing Na_2EDTA, 30 ml blood lysis (BL) buffer (155 mM NH_4Cl, pH 7.4; 10 mM $KHCO_3$; 1 mM Na_2EDTA) is added. After gentle mixing and incubation on ice for at least 1 h, the lysate is centrifuged at 3000 rpm for 15 min in a Beckman TJ-6 table-top centrifuge. When the pellet is still somewhat red, it is washed once with 10 ml BL buffer and incubated for another 15 min on ice. The supernatant is removed and the pellet is dissolved in 3 ml blood-nuclei lysis (BNL) buffer (10 mM Tris-HCl, pH 8.2; 400 mM NaCl; 2 mM Na_2EDTA), after which proteinase K (final concentration 0.5 mg/ml; stock: 20 mg/ml) and sodium dodecyl sulphate (SDS; final concentration 0.5%) are added. The solution is incubated overnight at 65°C under gentle shaking.

DNA is extracted by adding 1 vol of Tris-saturated phenol, gentle shaking for 10 min and centrifugation at 3000 rpm for 15 min. One volume of a 24:1 chloroform:isoamylalcohol solution is added to the water phase followed by shaking for another 10 min. After centrifugation at 3000 rpm for 15 min the water phase is removed and 1/10 vol 3 M Na-acetate (pH 4.9) is added. After mixing, 2 vol ethanol (96%) are added and the DNA is allowed to precipitate for at least 15 min at room temperature. The DNA can be removed from the solution using a Pasteur pipette or, alternatively, by centrifugation at 3000 rpm for 15 min. The pellet is washed twice with 70% ethanol and dissolved in TE–4

3.2.3.2   First dimension separation

Equipment. For the 1D electrophoresis we use the small version of the apparatus (Fig. 3.1a). The small glass plates ($170 \times 195 \times 4$ mm) are cleaned, first with soap and subsequently with ethanol and acetone. A gel-holding cassette is made with 0.75 mm thick Teflon spacers separating the glass plates. The gel-holding cassette is fixed in the electrophoresis apparatus by clamping with Teflon screws. A plug (120 ml) to seal the bottom of the gel-holding cassette is poured from 1.5% agarose in $0.5 \times$ TAE (0.02 M Tris; 0.01 M Na-acetate; 0.5 mM Na_2EDTA, pH 8.0 by addition of acetic acid) in the bottom trough.

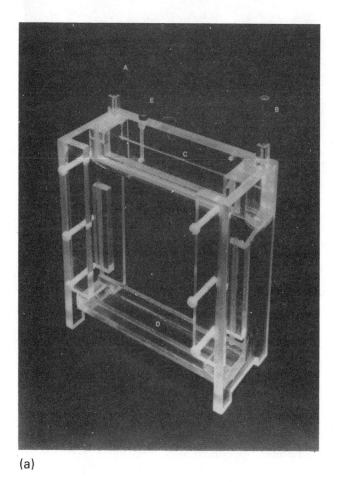

(a)

Fig. 3.1. Two versions of the gel apparatus used to run (a) 1D and (b) 2D gels: A, cathode electrode connection; B, anode electrode connection; C, platinum wire sunk in the back of the cathode buffer chamber; D, bottom trough under which the platinum wire, connected to B, is attached; E, cathode buffer chamber inlet (from Uitterlinden *et al.*, 1991b).

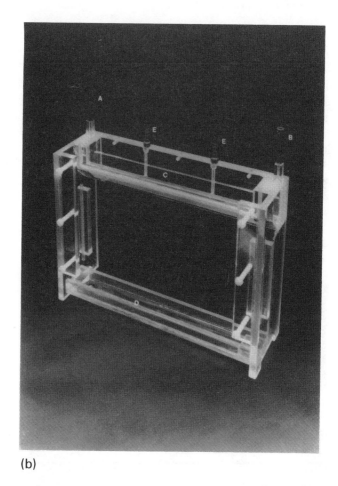

(b)

Fig. 3.1. *(continued)* Two versions of the gel apparatus used to run (a) 1D and (b) 2D gels: A, cathode electrode connection; B, anode electrode connection; C, platinum wire sunk in the back of the cathode buffer chamber; D, bottom trough under which the platinum wire, connected to B, is attached; E, cathode buffer chamber inlet.

Gel pouring. Gels are poured by pipetting 30 ml of 6% neutral acrylamide solution (acrylamide:bisacrylamide = 37.5:1 in 0.5 × TAE, kept at 4°C) between the glass plates. To allow polymerization, ammonium persulphate (APS; 200 µl of a 20% stock solution per 30 ml gel solution) and TEMED (30 µl per 30 ml gel solution) are added just before gel pouring. Immediately after gel pouring the comb is inserted and the gel is allowed to polymerize for at least 1 h at room temperature. The comb is removed and the wells are rinsed with 0.5 × TAE. The Teflon screws are loosened a little bit, so that tension is removed from the glass plates in order to prevent breakage when placed in a hot solution,

and the electrophoresis apparatus is placed in an aquarium containing 15 l
0.5 × TAE buffer kept at 50°C. We use the BioRad model 200/2.0 as power
supply. Alternatively, it is now possible to use the INGENY Jesse 1 power
supply, which is also suitable for blotting (see section 3.2.4). While loading the
DNA samples (slot volume in a 0.75 mm thick gel is approximately 15–25 µl),
the circulation pump should be switched off and a low voltage (10–20 V) can
be applied to prevent diffusion of the samples loaded in the wells.

Electrophoresis. Electrophoresis is carried out at 200 V for 3 h, during the first
5 min with the circulation pump switched off to allow the DNA fragments to
enter the gel. During electrophoresis, amperage should be about 0.10–0.15 A,
depending on buffer flow and level in the aquarium. Buffer level can decrease
owing to evaporation of water during electrophoresis. This can, however,
effectively be prevented by using polystyrene balls which completely cover the
buffer surface. After 3 hours the bromophenol blue dye marker has run off the
gel and the xylene cyanol dye marker is at the bottom edge of the gel.

Post-electrophoresis treatment and preparation of 1D lanes. After electrophoresis
the gel is removed from between the glass plates and stained for 15 min under
gentle shaking in reverse osmosis water (Milli-RO PLUS system, Millipore)
containing 4 µg/ml ethidium bromide, followed by 10 min destaining in RO
water. The gel is then placed on a UV trans-illuminator box (UVP products)
and examined under 312 nm UV light. The exposure of the 1D gel to UV light
should be as short as possible; exposures longer than 10 s will induce
single-strand breaks (see also 3.2.1.5). (Note that the UV flux of UV illuminators
decreases by a factor of 1.5–2 per year.) To protect the gel we keep a glass
plate between the gel and the UV box. This reduces the UV flux by a factor
of 10. Excision of the 1D lanes is done quickly with a Plexiglass or metal ruler
with a sharp edge. The range of fragments usually taken for 2D separation is
400–4000 bp (about 11.5 cm). About 4 kb is the upper limit of the resolution
of the 6% polyacrylamide gel.

3.2.3.3 Second dimension separation

Equipment. For the 2D separations we use the wider apparatus (see Fig. 3.1b),
which can easily accommodate two 1D lanes horizontally placed on top of the
wide 2D gel. Preparation of the glass plate cassette and pouring of the plug are
done as described for the first dimension separation.

Gel pouring. Standard denaturing gradient gels for 2D separation of micro- and
minisatellites consist of a 6% polyacrylamide gel containing a linear gradient of
10% stock solution of denaturants at the top and 75% at the bottom (100% =
7 M urea (Gibco-BRL Ultrapure) and 40% (by volume) formamide (Merck;
deionized by incubation with Bio-Rad AG501X8 mixed bed resin). These stocks
are abbreviated as 6A/10UF and 6A/75UF, respectively.

First, a small acryl plug is made by pouring 5 ml 6A/75UF solution, to which 40 µl APS (20% w/v stock solution) and 5 µl TEMED have been added for a quick polymerization, between the glass plates. This is necessary since otherwise the relatively heavy denaturing gradient gel will press water out of the agarose plug, which will distort the gradient. Linear gradients are poured using a computer-controlled gradient mixer from BIORAD, set at 8.3 ml/min for 5 min. Boundary solutions are then 25 ml 6A/10UF and 25 ml 6A/75UF, to which 170 µl APS (20% w/v stock solution) and 14 µl TEMED have been added. Alternatively, it is possible to use more simple and cheaper gradient makers provided, for example, by INGENY BV (Leiden, The Netherlands), but also by Gibco-BRL, Pharmacia and several other suppliers.

The gradient gel is poured from the 6A/75UF reservoir through a tube, held between the glass plates at the top. The top of the gel, which should be 5 mm below the top edge of the inner glass plate, is overlayered with 0.1% SDS to secure a straight surface of the gel. The gel is allowed to polymerize for at least 1 h. For rhomb gels (see 3.3.1), the direction of the gradient is diagonal to the first-dimensional separation lane. For this purpose gradients are poured into a gel apparatus tilted 45° sideways and with an extra rubber spacer placed at the top between the glass plates.

Electrophoresis. After polymerization, the 1D lanes are applied between the glass plates, as close as possible to the surface of the second dimension gel. Care should be taken not to include air bubbles between the 1D lanes and the denaturing gradient gel. To avoid floatage of the 1D lanes, they are sealed to the 2D gel and to the glass plates with hot 1% agarose in 0.5 × TAE solution. Then the apparatus is placed in the aquarium and the pump and electrodes are connected. The 2D gel is run at 60 °C for 13.5 h at 200 V. Under these conditions the amperage should be between 0.30 and 0.35 A. However, the amperage is highly dependent on the circulation speed of the pump and the buffer level in the tank: the higher the pump speed, the higher the amperage; the lower the buffer level in the tank, the lower the amperage.

Post-electrophoresis treatment of a 2D gel. After electrophoresis the 2D gel is stained/destained and examined under 312 nm UV as described previously. After documentation the gel is exposed to 312 nm UV for about 4 min to fragment the DNA. Fragmentation supports an efficient transfer of the DNA during the electrotransfer procedure. Note that the exposure time is dependent on the type of UV instrument used which also shows a decrease in UV flux in time. Therefore, optimal UV exposure time should be determined experimentally. For subsequent manipulation of the gel a piece of dry filter paper is placed on the gel. The gel will stick to the paper and can now be easily removed and subsequently released from the paper by wetting it again in buffer. Usually, the gel is rolled up in the paper (making it less prone to breakage) and allowed to equilibrate in 0.5 × TBE (89 mM Tris-borate, pH 8.0; 89 mM boric acid; 2 mM Na_2EDTA) for at least 20 min, but no longer than 2 h.

3.2.4   Electrotransfer of separation patterns

3.2.4.1   *Equipment*

The 2D polyacrylamide gel separation patterns are transferred to a nylon membrane using a home-made semi-dry electrotransfer instrument, consisting of a stainless steel cathode plate and a titanium-ceramic-coated anode plate (Fig.3.2). The plates are mounted in Perspex with the stainless steel cathode plate acting as lid. The plates are 200 × 400 mm in size and can easily accommodate multiple gel stacks on top of each other (see below). It is our experience that this instrument, which is now commerically available through INGENY BV (Leiden, The Netherlands), results in more efficient transfer than comparable commercially available instruments (e.g. BIORAD, Biometra).

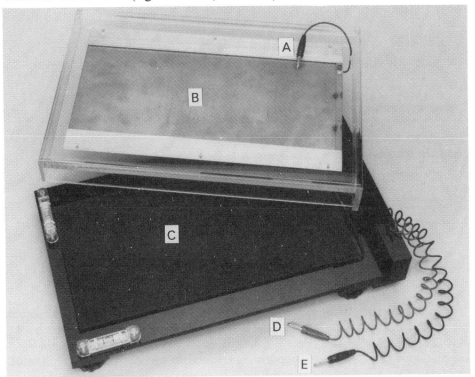

Fig. 3.2. Electrotransfer apparatus based on one stainless steel and one titanium ceramic coated plate: A, cathode electrode connection which is directly attached to the stainless steel cathode plate B; C, ceramic anode plate; D, anode electrode; E, cathode electrode.

3.2.4.2 Procedure

After equilibration in 0.5 × TBE the gel is removed from the filter paper and placed on five sheets of wetted Whatmann 3 MM paper on a used X-ray film. The nylon hybridization membrane (Hybond N+, Amersham, or other comparable positively charged nylon membranes) is placed over the gel with five sheets of

Whatmann paper on top. It is very important to remove air bubbles between the different layers (gel, membrane and filter paper). This is done by rolling a 10 ml pipette with some pressure over the transfer stack every time a new layer has been added. Finally, the whole transfer stack is turned upside down and placed on the anode plate of the electrotransfer instrument. The membrane is now below the gel.

The anode plate with the transfer stack(s) is placed inside the transfer apparatus and the lid, containing the cathode plate, is put on top. Electrotransfer is performed for 1.5 h at a constant amperage of 400 mA. During this period the voltage increases from 5–7 V to 12–30 V, depending on the number and size of the gels between the plates. We routinely process multiple (at least four) stacks simultaneously on top of each other (Fig. 3.3 schematically depicts a two-gel stack).

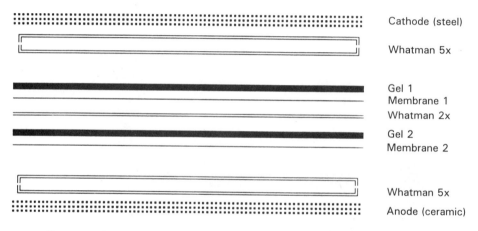

Cathode (steel)

Whatman 5x

Gel 1
Membrane 1
Whatman 2x

Gel 2
Membrane 2

Whatman 5x
Anode (ceramic)

Fig. 3.3. Schematic depiction of a two-gel stack used in electroblotting 2D DNA typing gels.

Immediately after transfer, the stacks are dismantled and, in order to denature the DNA fragments, the membranes are incubated for at least 2 h in 0.5 M NaOH, 1.5 M NaCl. The relatively long denaturation period and the high NaCl concentration are essential to suppress background in the subsequent hybridization analysis. The membrane is neutralized by rinsing twice in large volumes of $2.5 \times$ SSC. To crosslink the DNA fragments to the membrane, the DNA side is exposed to 312 nm UV light for 45 s. Note that the time of this UV cross linking procedure should be determined empirically and depends on the UV-flux of the apparatus which can decrease in time.

3.2.5   Hybridization analysis

In principle, any type of probe can be used in hybridization analysis of 2D DNA typing patterns which is performed essentially as according to Church and Gilbert (1984). Here we will only describe the use of micro- and minisatellite core probes. Because of the advantageous hybridization kinetics of their repeat elements good signals are obtained with such probes in hybridization analysis. In general only 2 h of hybridization time and a few hours of exposure are required (see Fig. 3.4).

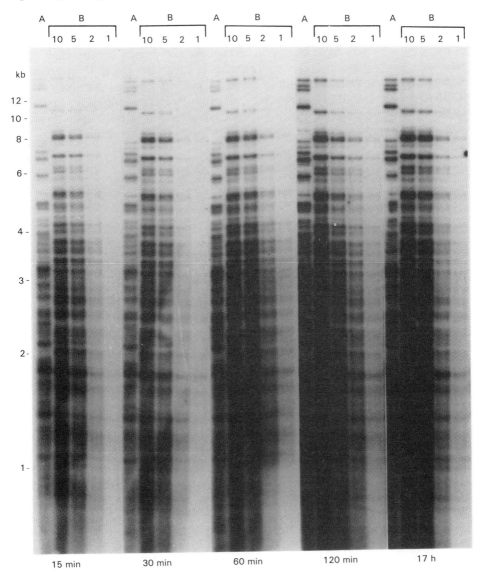

Fig. 3.4. Time course hybridization experiment. Southern blots containing human Hae III digested genomic DNA of two unrelated individuals (A and B) were subjected to 15 min, 30 min, 60 min, 120 min and 17 h of hybridization with minisatellite core probe 33.15. Washing conditions and exposure times were identical. Of individual B, 10, 5, 2 and 1 μg of (digested) DNA were applied to the gel; of individual A, 10 μg was applied.

3.2.5.1 Preparation and labelling of the probes

The micro- and minisatellite polymer probes used for two-dimensional DNA typing are a mixture of synthetic, double-stranded tandem repeat motifs with an average length between 1000 and 10.000 bp. Fig. 3.5 schematically depicts how these probes are prepared. This can be done by including both a ligation and a PCR step or by only including PCR for polymerization of oligonucleotides. (Uitterlinden *et al.*, 1989a; Ali and Wallace, 1989; Vergnaud, 1989; Vergnaud *et al.*, 1991a). Both protocols result in satisfactory probes but in view of its ease the PCR protocol is now preferrred. Table 3.1 lists a selection of oligonucleotide sequences routinely used as core probes in 2D DNA typing (Uitterlinden *et al.*, 1991). However, it is likely that many other oligonucleotide sequences are also suitable. Fig. 3.6 shows examples of 2D DNA typing patterns obtained with human genomic DNA using different micro- and minisatellite core probes.

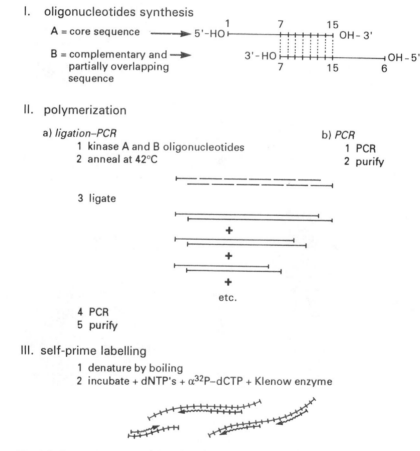

Fig. 3.5. Schematic representation of the preparation and labelling of synthetic, double-stranded polymers of tandem repeat motifs, used as probes.

Table 3.1. Core probes selected for 2D DNA typing[a]

Name	Sequence
HBV2	GGACTGGGAGGAGTTGGGGG
ZETA	TGGGGCACAGGCTGTGAG
33.15	AGAGGTGGGCAGGTGG
HBV1	GGAGTTGGGGGAGGAG
HBV3	GGTGAAGCACAGGTG
HBV4	GAGAGGGGTGTAGAG
HBV5	GGTGTAGAGAGGGGT
YNZ22	CTCTGGGTGTCGTGC
M13	GAGGGTGGCGGCTCT
INS	ACAGGGGTGTGGGG
HYCO	GCTGGTGGGCAG
33.6	AGGGCTGGAGG
ENHA	GCTGTGGTTT
RECO	CAGGTGG
TELO	TTAGGG
BkmC	GACA
BkmT	GATA
$(CAC)_n$	CAC
$(AGC)_n$	AGC
$(TCC)_n$	TCC

[a]The sequence shown represents the repeat unit. Each probe consists of multimers of the form (sequence)$_n$. A more comprehensive list is provided in Uitterlinden *et al.* (1991a). From that list, however, not all probes have been extensively tested.

The desired oligonucleotide core probes are prepared as follows. For each core probe two partially complementary and overlapping oligonucleotides are synthesized using the Gene Assembler Plus (Pharmacia) and purified by FPLC. If unpurified oligonucleotide is used we have observed inhibition of the ligation reaction. About 500 ng of each oligonucleotide are mixed and subjected to PCR amplification. Besides the oligonucleotides the PCR reaction mixture should contain 50 mM KCl, 10 mM Tris-HCl (pH 8.0), 1.5 mM $MgCl_2$, 200 µM dATP, 200 µM dTTP, 200 µM dCTP, 200 µM dGTP and 1 unit of Tth polymerase (HT Biotechnology Ltd, UK) in a final volume of 100 µl. The reaction mixture is subjected to 2×30 cycles consisting of 30 s denaturation at 92°C, 30 s annealing at 55°C, and 2 min extension at 72°C in a BIOMED thermal cycler (Braun, Germany). The probes are subsequently purified by phenol extraction

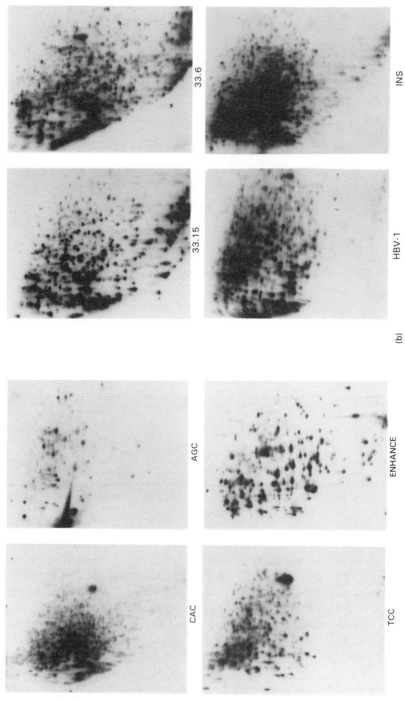

Fig. 3.6. Examples of 2D DNA typing patterns obtained with human Hae III-digested genomic DNA using different (a) microsatellite and (b) minisatellite core probes. The same filter was re-hybridized with the different probes.

and ethanol precipitation. They are generally between 1000 and 10 000 bp in length.

For use in hybridization reactions 2 ng of the probes is [α^{32}P]dCTP labelled, by the random-primed oligolabelling method (Feinberg and Vogelstein, 1983, 1984), after boiling for 5 min and reannealing at room temperature in the presence of 1 unit Klenow enzyme (BRL), 2 µM dNTP; 50 mM Tris-HCl, (pH 7.2) and 10 mM MgCl$_2$. Specific activities of 3×10^9 cpm/µg can be obtained. It is also possible to apply non-radioactive probe labelling protocols. Although we have had some success with kits from Boehringer, a routinely applicable protocol has yet to be established.

3.2.5.2 *Hybridization*
Hybridization reactions are carried out at 65°C in a hybridization oven (GFL) with rotating glass tubes. Filters are briefly pre-wetted in 2.5 × SSC and put into the glass tubes. For pre-hybridization, 10–20 ml of the pre-hybridization mix (7% SDS; 0.5 M phosphate buffer, pH 7.2; 1 mM Na$_2$EDTA) is added to the filter in the hybridization tube. Filters are pre-hybridized for 15 min at 65°C. In one tube up to 10 filters can be hybridized together provided 1 labelled probe is used per 2 standard size 2D filters (18 × 28 cm each). The probe, made single stranded by boiling for 5 min, is subsequently added to the pre-hybridization mix in the tube. For polymer probes (longer than 1000 bp) a 2 h hybridization period is sufficient (see also Fig. 3.3).

3.2.5.3 *Washing and autoradiography*
Filters are normally washed twice for 30 min under relaxed conditions in 2.5 × SSC, 0.1% SDS at 65°C. For higher stringencies an additional wash for 30 min in 1.0 × SSC, 0.1% SDS at 65°C is applied. Autoradiography is performed between intensifying screens by exposing Kodak XAR film for 2–72 h at −80°C.

3.2.5.4 *Stripping of filters*
To remove bound radioactive probe, the hybridization membranes are stripped in 400 ml 0.1 × SSC, 10% SDS at 100°C in a metal container. SDS is only added to the SSC when this solution is at 100°C. After cooling down to room temperature the filter is washed for an additional 5 min in fresh 2.5 × SSC.

3.3 EXPERIMENTAL CONSIDERATIONS
3.3.1 Denaturing gradient gel constellation
The initial two-dimensional DNA separation experiments were performed using linear denaturing gradients and a perpendicular orientation of the two dimensions, according to the original system as described by Fischer and Lerman (1979a and b). A problem frequently observed with these standard separation patterns is the condensation of spots in the upper left part of the 2D pattern (high molecular weight range and low denaturant concentration). As shown in Fig. 3.7, different

2D gel constellations are possible, which do in some cases result in improved resolution (Fig. 3.8). Improved resolution was obtained when, instead of a linear denaturing gradient, an exponential gradient was applied with smaller increments of denaturants concentration at the top of the gel and larger increments at the bottom. To improve resolution of large restriction fragments and resolution in the top of the gradient we designed so-called rhomb-exponential gradients (REX gels). In this gel type the exponential denaturing gradient is oriented diagonally to the direction of electrophoresis and first dimension separation pattern. Thus, large restriction fragments can travel further into the gel and small restriction fragments are more evenly distributed over the denaturing gradient, under electrophoretic conditions identical to those for standard gels.

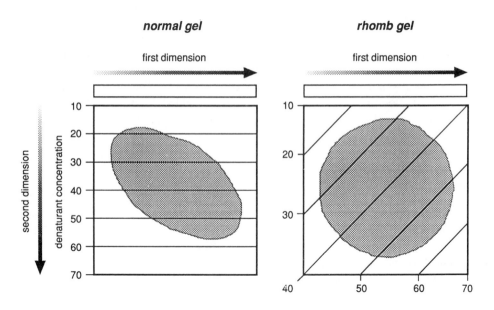

Fig. 3.7. Schematic representation of different denaturing gradient gel constellations showing the predicted influence of variations in the shape and direction of the denaturing gradient on the separation pattern.

3.3.2 Buffer dependency of separation patterns

3.3.2.1 Ion concentration
Ion concentration in the electrophoresis buffer appeared to influence critically 2D DNA separation patterns. For example, when $0.5 \times$ TAE instead of $1.0 \times$ TAE is used as the buffer present in the gel and in the circulation buffer, a separation pattern is observed in which the spots are spread over a wider area of the gel,

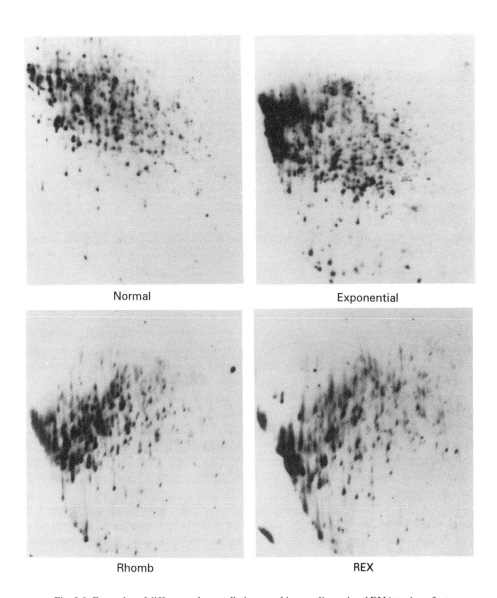

Fig. 3.8. Examples of different gel constellations used in two-dimensional DNA typing of rat genomic DNA digested with Hae III and hybridized with minisatellite core probe 33.15. Identical first dimension separations (in a 6% PAA gel) were subjected to different second dimension conditions. 'Normal' stands for a linear 10–75% denaturant concentration contained within a 6% PAA gel; 'Exponential' denotes the application of an exponential 10–75% denaturant concentration in a 6% PAA gel; 'Rhomb' is a linear 10–75% denaturant concentration in a diagonal orientation to the first dimension separation, and REX is an exponential gradient of the 10–75% denaturants concentration in a diagonal orientation to the first dimension (from Uitterlinden *et al.*, 1991a).

both in the first and in the second dimension electrophoresis. This was found with human, cattle and rodent DNA.

3.3.2.2 Water quality

A note of warning should be made here with respect to the influence of water quality, used to prepare the buffers, etc., on the separation patterns obtained in 2D DNA typing. Because of the presence of some hitherto unknown contaminating substances in water of particular purity (especially of water prepared by condensation) the separation pattern can be negatively influenced. We therefore routinely use water purified by reverse osmosis (Milli-RO Plus, Millipore).

3.3.3   Reproducibility

The reproducibility of 2D DNA typing is an essential prerequisite for the routine application of this electrophoretic analysis system. Sources of experimental variation are listed in Table 3.2. As can be derived from this list, the complicated procedure of 2D DNA typing with its various steps can give rise to ample variations in the spot patterns eventually obtained. A logical way to circumvent some of these sources of variations is to run as many gels as possible under exactly the same conditions, thereby avoiding manual interference as much as possible. We are at present involved in the development of a semiautomatic instrument to run the first and second dimension in one gel without any handling required (IngenyMapper; INGENY BV, Leiden, The Netherlands). In this system multiple gels can be run simultaneously (Mullaart *et al.*, 1993). Below, we provide estimates of the reproducibility of 2D DNA typing with the current protocols.

Fig. 3.9 shows examples of intra- and intergel reproducibility of 2D DNA typing gels of rat genomic DNA probed with minisatellite core probe 33.15. As can be seen, the patterns are sufficiently reproducible to allow sample comparison. However, particular spots are observed (typically less than 1% of the total number of spots) which are sensitive to small electrophoretic variations in the second dimension. Thus far we have no explanation for the cause of this phenomenon.

The use of marker DNA fragments facilitates 2D gel comparisons. The marker points serve to adjust for gel distortions and differences between electrophoretic runs. We routinely mix the DNA digests to be analysed with sets of restriction fragments of lambda DNA. After co-electrophoresis, these markers can be visualized by ethidium bromide staining and by hybridization with labelled lambda DNA after or before the 2D separation patterns have been probed with the core sequences. One such marker set is shown in Fig. 3.10 (see also Chapter 4; Fig. 4.2). So far we have not seen any cross-hybridization of lambda restriction fragments with micro- or minisatellite core probes. Table 3.3 gives the deviations observed in the positions of the lambda spots. These results were obtained by repeated analyses of the same DNA sample and, therefore, provide a measure of the experimental error in spot position (see also Fig. 3.10). The somewhat higher experimental error after hybridization analysis as compared with

Table 3.2. Sources of variation in 2D DNA typing

Stage	Step	Artefact	Possible cause
Sample preparation	DNA isolation	Vertical streaks in pattern Horizontal streaks in pattern	Single-strand breaks Double-strand breaks
	Enzyme digestion and loading	HMW background Different numbers of spots	Partials Different amounts loaded on gel
Electrophoresis	First dimension	Double spots Variation in position	Band smiling Electrophoresis variables: time PAA concentration buffer concentration
	Second dimension	Vertical streaks Variation in x axis position Variation in y axis position	UV exposure of 1D lane (single-strand breaks) 1D lane application (stretching) Gradient constellation Electrophoresis variables: temperature time PAA concentration buffer concentration
	Electrotransfer	No transfer of large fragments Holes in spot pattern Less spots visible	UV exposure (for fragmentation of large fragments) to short Air-bubbles in stack Denaturation variables: buffer concentration time
Hybridization analysis	Hybridization	Fewer spots visible	Core probe quality: average size labelling Stringency of hybridization/washing
	Autoradiography	Spots blurry	Exposure at −80°C (instead of room temperature)

(a)

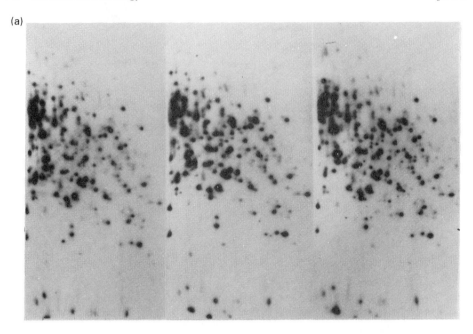

(b)

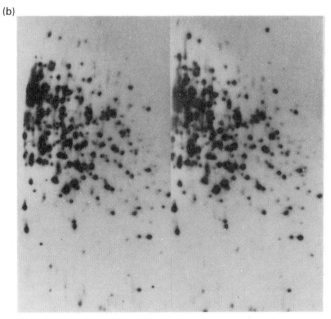

Fig. 3.9. Intra- and inter gel reproducibility of 2D DNA typing patterns of Hae III digested rat genomic DNA probed with minisatellite core probe 33.15. (a) The upper panel represents a single gel containing three adjacent separation patterns of the same DNA sample. (b) The lower panel represents an independent gel showing two adjacent separation patterns of the same DNA sample run in an independent experiment (from Uitterlinden *et al.*, 1991a).

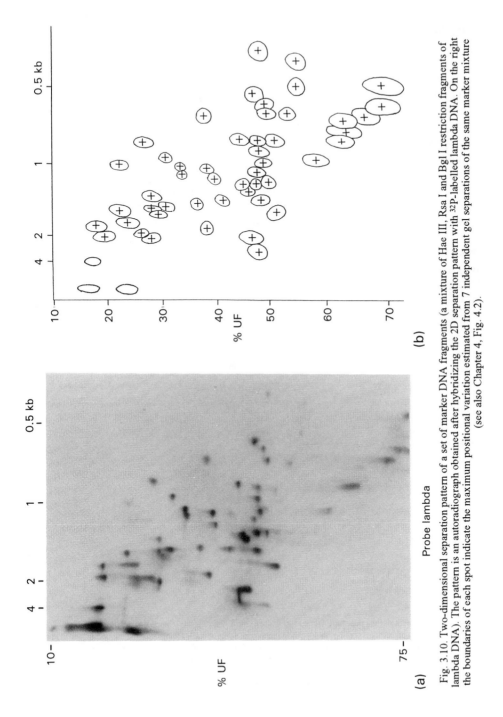

Fig. 3.10. Two-dimensional separation pattern of a set of marker DNA fragments (a mixture of Hae III, Rsa I and Bgl I restriction fragments of lambda DNA). The pattern is an autoradiograph obtained after hybridizing the 2D separation pattern with ^{32}P-labelled lambda DNA. On the right the boundaries of each spot indicate the maximum positional variation estimated from 7 independent gel separations of the same marker mixture (see also Chapter 4, Fig. 4.2).

ethidium bromide staining is due to some stretching of the gel during blotting. Using the same series of gels the experimental error in the position of spots in the patterns obtained with minisatellite core probe 33.15 was assessed. The results in Table 3.3 indicate errors in spot position virtually identical to the errors found with the marker spots.

Another potential source of error is the lack of reoccurrence of a spot known to be present in a given DNA sample and the occurrence of spots that should not be present but are due to dirt or other forms of contamination. With respect to the former, it was found (from the same set of repeated analyses) that on average 1 out of 100 spots cannot be reproduced. Spots as a result of, for example, contamination with dirt particles are virtually never observed.

Finally, it should be noted that considerable variation can occur in the total number of spots detected. Naturally, fewer and fewer spots are detected when the stringency of hybridization and washing conditions is raised (less SDS and lower salt concentrations). This has already been discussed in Chapter 2 (2.2.4.1). Under sub-optimal experimental conditions (see Table 3.2) the total number of spots can decrease. In general, under optimal conditions one can expect to find a minimum number of spots to be reproducibly detected.

Table 3.3. Reproducibility of 2D spot positions[a]

	Size	% UF
	X-coordinate (mm)	Y-coordinate (mm)
A. LAMBDA MARKER SPOTS		
After ethidium-bromide staining		
Intra-gel[b]	0.60	0.47
Intra-gel[c]	1.18	1.93
After hybridization with lambda		
Intra-gel[b]	1.07	0.81
Intra-gel[c]	1.59	2.66
B. CORE PROBE 33.15 SPOTS[d]		
Intra-gel[b]	0.97	1.29
Intra-gel[c]	1.32	2.38

[a]Based on comparing the position of spots in repeated 2D DNA typing experiments of the same genomic DNA sample mixed with lambda marker fragments. Comparison of ethidium-bromide stained gel patterns provides an estimate of the reproducibility of the electrophoretic run; comparison of lambda or core probe 33.15 hybridization patterns includes the reproducibility of the spot positions after transfer of the separation pattern to the nylon membrane.
[b]Seven pairwise comparisons of two samples run on the same gel.
[c]Comparison of seven spot patterns on seven replicate gels.
[d]Based on comparing 30 randomly chosen spots among 465 detected in total with core probe 33.15 in a human individual in a single hybridization experiment.

3.4   PATTERN EVALUATION BY IMAGE ANALYSIS

Two-dimensional DNA typing can provide spot patterns consisting of hundreds of spots. Such patterns can be analysed by eye but computerized image analysis may provide a time-saving and rapid alternative solution for the accurate evaluation of the information contained within a 2D DNA typing pattern. However, it should be realized that image analysis systems are not better than the eye. Indeed, it is our experience that for simple comparisons of spot patterns the eye is generally more rapid in finding differences than image analysis systems. Indeed, especially when the samples to be compared are run on one gel it is generally easy to find the variant spots quickly. For this purpose we cover the autoradiograph with the patterns to be compared (generally two) with a grid on a transparency, on a light table. This is illustrated by Fig. 3.11, in which a 2D DNA typing pattern of a tumour genome is compared with that of the corresponding normal genome (genomic DNA from white blood cells). The variant spots are easily scored and indicated on the transparency.

However, for more complicated tasks, such as the analysis of co-segregation of spots in large pedigrees (see Chapter 4), image analysis systems coupled to genetic analysis programs, for example for detecting linkage, are almost indispensable. Currently there are several image analysis systems commercially available that can handle two-dimensional gel patterns. Advanced systems are the Tycho system (N. Anderson *et al.,* 1981; N. Anderson, 1991) and the Quest system (Garrels, 1989), but many other systems are now available (e.g. Bio Image system, Millipore; Masterscan, Scanalytics). More recently, we have developed a system specifically for the interpretation of 2D DNA typing patterns, the so-called IngenyVision system (INGENY BV, Leiden, The Netherlands; te Meerman *et al.,* in preparation). These and other commercially available programs all perform the following steps: (1) image capturing; (2) spot detection; (3) calibration; (4) matching; (5) data base generation and searching.

3.4.1   Image capturing

Images can be captured from autoradiographs or directly from radioactive hybridization filters by means of phosphor storage screens (available from, for example, Molecular Dynamics, Fuji, BioRad). The latter have the advantage that the dynamic range of signal is much larger, which allows quantitative determinations. Autoradiographic images can be acquired by means of scanners or cameras, where there is a trade-off between speed of registration (cameras are faster) and resolution (scanners have a higher resolution). In general, image capturing is a simple and accurate process and not a limiting factor in computerized image analysis.

3.4.2   Spot detection

Spot detection involves the recognition of spots as such and their definition in x–y coordinates. This necessitates local background subtraction, the removal of

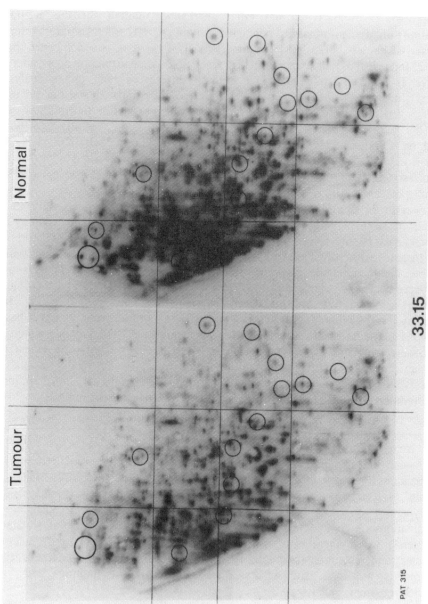

Fig. 3.11. Comparison of 2D DNA typing patterns of a breast tumour genome with the corresponding normal genome of white blood cells. The probe used was 33.15. Potential spot differences can be indicated on the transparency overlay by circles.

streaks and the recognition of artefactual spots as such. Basically, spot detection can be accomplished by the use of so-called threshold routines. That is, spots are detected on the basis of their increased intensity above a certain threshold. A refinement, allowing the more difficult tasks of recognizing spots in clusters, streaks or artefactual spots can be accomplished, for example, by the Gaussian fitting procedure provided in the Tycho/Kepler system (N. Anderson *et al.,* 1981). Especially when spot patterns are dense with high local background and spots of different morphology and intensity, spot detection systems must be very good to match the eye.

3.4.3 Calibration

In principle, calibration is nothing more than assigning x–y values to each spot (fragment size and melting temperature). Most commercially available image analysis systems use a linear calibration in x–y dimensions. However, this does not allow correction for intergel variations. In these systems this correction process is performed during matching (see below). Correction is absolutely necessary because the positions of individual spots and/or groups of spots may differ considerably from gel to gel. Such deviations, which to some extent can be avoided by standardizing the experimental conditions and/or running multiple samples on the same gel, find their origin in the various sources of experimental variation in 2D DNA typing patterns (Table 3.2). Since in 2D DNA typing each gel gives rise to multiple (hybridization) patterns, calibration for one pattern also provides the formula for all other patterns derived from this gel.

Alternatively, calibration can be done in a non-linear fashion by using a set of internal marker spots corresponding to fragments of known size and melting temperature (derived either from a known DNA molecule such as phage lambda DNA or from constant spots in hybridization patterns). The presence of constant spots in different 2D separation patterns allows exact calibration of unknown spots and comparison with other 2D separation patterns.

We have recently developed an image analysis system (IngenyVision; te Meerman *et al.,* in preparation), in which the calibration step is based on the use of internal marker fragments (e.g. phage lambda digests, plasmid digests). This system relies on the use of built-in databases (see below) of spot patterns, corresponding to different core probes and a set of marker fragments. Since the set of markers is always mixed with the genomic DNA sample to be analysed, each gel is defined by its own marker pattern. It is this marker pattern that is used as an internal standard for the transformation of first the reference marker database and then the reference core probe databases (illustrated in Fig. 3.12). These transformed databases can then be used to score for the presence or absence of particular spots or groups of spots in the patterns corresponding to the sample. Spots in the sample patterns should now perfectly fit the corresponding spots in the transformed databases, provided that the marker fragments accurately follow deviations in the position of core probe spots. Thus far we have no indication that this is not the case.

It is almost needless to say that with dense patterns in such marker-based systems spot pattern calibration should be extremely accurate to prevent mismatching of spots. In practice, it is necessary to use interactive systems allowing corrections to be made. A good strategy, for example, is the use of a multiwindow system (widely applied in image analysis systems) to compare systematically the same part of a pattern for different individuals. For example, in studying pedigrees the INGENYVision system allows to observe simultaneously the same part of the 2D pattern in multiple individuals. This allows the spot detection and calibration processes to be checked by eye in a systematic way. This strategy offers the advantage that it is both rapid (as compared with an analysis entirely by eye) and accurate (as compared with fully automatic analysis).

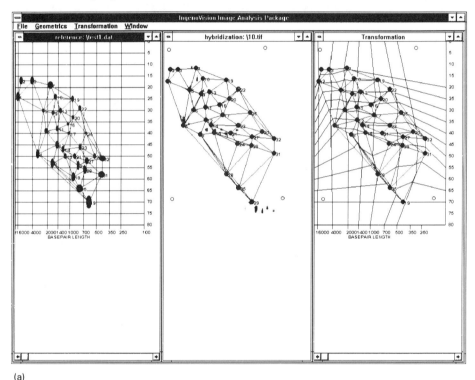

(a)

Fig. 3.12. The INGENYVision system. (a) Calibration step. A reference marker set (spot pattern on the left) is compared with the same marker set added as an internal standard to the genomic DNA samples to be analysed (pattern in the middle). The middle pattern shows the actual marker hybridization pattern of a particular gel with a genomic DNA separation pattern. The marker spots corresponding with those present in the reference marker set are indicated (numbers are arbitrary) and connected by lines to form triangles.Using triangle-transformation routines the reference pattern is transformed to match the internal standard of this particular gel. The resulting transformed reference pattern is shown on the right. This figure is also provided in colour as a frontispiece to the book. The marker spots are indicated in green and the triangles in red. The yellow points in the corner are used for alignment of the marker pattern and the different core probe patterns obtained with a given gel.

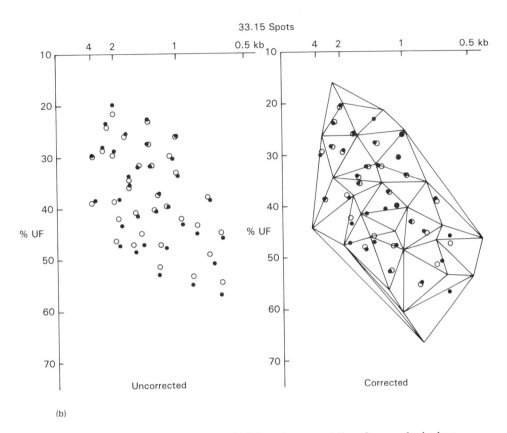

Fig. 3.12. The INGENYVision system. (b) Schematic representation of core-probe database comparisons. On the basis of the transformation formula obtained for the marker set, the other reference patterns (one for each micro- and minisatellite probe) are transformed. The spots they contain should now match the corresponding spots in the sample hybridization patterns. Owing to a limited number of marker spots in some areas perfect matching for all spots was not obtained in this particular case.

3.4.4 Matching

When calibration is only linear, correction for intergel variations should take place during matching. In most image analysis systems matching occurs by the selection of so-called landmark spots. These are clearly visible spots shared by the patterns to be compared. With these landmark spots as anchor points the patterns are transformed, to one another or to a reference pattern. Also here mismatching is a frequent phenomenon, especially when the spot patterns are dense. Hence, an interactive system is desirable.

With non-linear calibration on the basis of marker fragments matching of spot patterns is relatively simple and a matter of scoring the presence or absence of spots. In addition, matching can be done on the basis of several other criteria (e.g. intensity, morphology) which can be used in a particular order to define subsets of differences and to define limits to declare a match (e.g. to prevent a large and intense spot from being automatically matched to a very faint and small spot at an identical position in the pattern). A likely hierarchical order in this respect is matching on the basis of (a) position, (b) area, and (c) intensity.

3.4.5   Database generation, searching and matching

Databases for 2D DNA typing patterns can be generated containing information on spots according to the following scheme:

Sample information:
1   Species
2   Family or group
3   Individual number

Method information:
1   Restriction enzyme
2   Gel modality
3   Probe

Spot information:
1   Length in basepairs
2   Melting temperature
3   Intensity
4   Morphology
5   Chromosomal location
6   Linkage group
7   Locus identity
8   Number of alleles
9   Allele frequency
10   Mutation frequency

For 2D DNA typing patterns blotting membranes can be hybridized to different probes each of which generates a given number of spots in an individual of a certain species. Some of these spots are discarded, for example, because they are in one of the gel areas difficult to interpret, because they represent constant rather than polymorphic loci etc.

At present we are involved in the generation of databases of spots suitable for linkage analysis. For this purpose we are studying co-segregation of spots in the CEPH panel (see also Chapters 2 and 4), using a set of suitable micro- and minisatellite core probes. In this study we make use of the many locus-specific markers with known segregation patterns in the CEPH panel. Co-segregation of spots with one of these markers allows us to assign chromosomal positions to the loci corresponding with particular spots. In this way, using our INGENYVision image analysis system, databases are being generated corresponding to sets of spots representing useful markers for linkage analysis. After correction for overlap among the different core probes, each database will define a unique set of loci detected by a given core probe.

On the basis of the characteristics mentioned above databases can then be searched for the presence or absence of one or more particular spots. Each search creates a data output file, which can be entered into other programs, for example, genetic linkage programs. In combination with the analysis of the CEPH reference set of pedigrees this allows for the construction of a genetic map of the human genome based on 2D spot variants (see also Chapter 4).

Finally, communication can be set up with other databases, for example, to achieve on-line identification of the position of disease-related genetic loci in the human genome. These databases have already been initiated in the framework of the Human Genome Project such as the database containing information on genetic data of genomic loci (Genome Database, GDB; Johns Hopkins, Baltimore, MD, USA) which on its turn is coupled to databases containing information on sequence of the locus and its variants (Genbank, USA, and EMBL, Europe). In addition, many related databases are being set up now such as those for collecting physical mapping information on particular chromosomes (Los Alamos database for chromosome 19; ICRF database for several human chromosomes) and even the complete human genome in the form of YACs (YAC libraries from University of Washington, CEPH in Paris and ICRF in London). In parallel, international efforts have been initiated to compile a genetic linkage map of the human genome by analysis of an ever-growing set of genetic markers on a given set of pedigrees by several laboratories (NIH/CEPH collaborative mapping group, 1992; Weissenbach *et al.*, 1992). It can be envisaged that each spot in a 2D pattern will contribute to such linkage maps, thereby enhancing its information content.

4

Applications

4.1 INTRODUCTION

Genome scanning by 2D DNA typing is applicable in the following areas: (1) classification, diagnosis and genetic monitoring of microorganisms; (2) mapping of genetic traits (including human disease genes) by linkage or association analysis; (3) the analysis of genomic instability in tumour progression and during ageing; (4) measuring mutation rates at various genomic sites. Eventually such studies will identify particular genomic regions as spot variants in a 2D spot pattern. By matching with a computer database or, alternatively, by direct isolation of the DNA fragment corresponding to the spot of interest information on the particular region can be obtained.

Among the various applications of 2D DNA typing genomic analysis of microorganisms stands out as the most straightforward one. Indeed, the relatively small size of the genomes of bacteria and lower eukaryotes, such as yeast, worms and flies, greatly simplifies detection in gel after electrophoretic resolution. Whereas for the large genomes of mammals hybridization analysis is necessary for the detection of repeat elements as anchor points in total genome scanning (see below), the small-sized genome of, for example, the bacterium *Escherichia coli*, allows direct detection of all fragments resulting from an Eco RI digest resolved in a 2D DNA typing gel simply by ethidium bromide staining (Fischer and Lerman, 1979a). For microorganisms with genomes of less than 10 million basepairs direct detection is usually possible, especially when gels larger than the standard type are used. For lower eukaryotes with genomes of more than 10 million basepairs hybridization analysis with repeat element probes is necessary. Alternatively, end labelling techniques and/or interrepeat PCR can be used (see below and Chapter 5). Genome analysis of lower eukaryotes may find applications in classification, diagnostics and monitoring industrially important cultures for genomic instabilities.

For complex DNA molecules of several billion basepairs such as those from higher eukaryotes, EtBr staining of 2D separation patterns of genomic restriction enzyme digests will result in an unresolvable blur corresponding to millions of

Electrophoresis

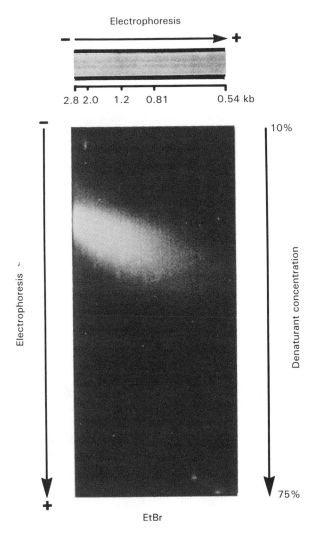

Fig. 4.1. Ethidium bromide stained 2D separation pattern of human genomic DNA. On top
the 1D separation pattern used for the second dimension separation is shown (from
Uitterlinden *et al.*, 1989a).

different restriction fragments (see Fig. 4.1). To obtain interpretable 2D spot
patterns, there are several possibilities for selectively visualizing particular groups
of sequences which are dispersed over the genome, e.g. hybridization analysis
using repetitive sequences as probes, interrepeat PCR (irPCR) using primers for
repetitive motifs, selective end labelling of restriction fragments. The information
to be obtained from such an analysis critically depends on the numbers and types
of sequences used as landmarks for a particular genome; the choice of probe,
primers and/or restriction enzymes determines the scanning efficiency. In this

chapter we will focus on the results that can be obtained either by direct detection (viral and bacterial genomes) or by hybridization analysis with repeat element probes.

4.2 GENOME ANALYSIS OF LOWER ORGANISMS

In view of the resolving capacity of the 2D DNA typing system as described here (which includes a separation in a denaturing gradient gel), different 'scanning efficiencies' can be obtained for genomes of different lower organisms. For viral DNA, usually the genomes do not exceed lengths of 200 kb and hence all restriction fragments of a given digest (typically less than 100) can be visualized by ethidium bromide staining. However, for mycoplasma and bacteria, genome sizes can be up to several million basepairs which will result in several hundreds to thousands of restriction fragments. Under particular circumstances (depending on the restriction enzyme, gel size, etc.) these may all be resolved in a single 2D gel. For lower eukaryotes, such as yeast, the number of fragments generated is too high to allow all fragments to be resolved on standard format gels and hence hybridization analysis with repeat elements as probes has to be applied. Still little is known of the genomic distribution of especially the micro- and minisatellite elements in these genomes and hence not much is known of the efficiency of the genome scans obtained from these organisms.

4.2.1 Analysis of viral and bacterial genomes
As an example of the 2D DNA typing of viral DNA the analysis of the bacteriophage lambda will be discussed. The genome of lambda is 48 502 bp long. Digestion with the restriction enzyme Eco RI results in the generation of six fragments. Fischer and Lerman (1979a, b) have demonstrated that these fragments show a length-independent separation in a 2D gel and that there is good correspondence between calculated melting temperatures of particular melting domains in the lambda genome and the actual position of the corresponding restriction fragments in the denaturing gradient gel (Fischer and Lerman, 1979a, b, 1983; Lerman et al., 1983). In Fig. 4.2a the melting map of the complete lambda genome is shown as derived from the computer algorithm developed by L.S. Lerman (see also Chapter 1). The map shows the presence in the lambda genome of several melting domains which will undergo cooperative melting and lead to the focusing of restriction fragments in the denaturing gradient gel. Underneath the melting map the positions of some of the larger Hae III, Rsa I and Bgl I restriction fragments are indicated.

In Fig. 4.2b the actual 2D gel separation pattern is shown of a mixture of digests of lambda DNA obtained with Hae III, Rsa I and Bgl I. For all fragments containing a melting domain we observed close correspondence between the predicted order of melting (resulting in a state of considerably reduced electrophoretic mobility in the denaturing gradient gel) and the position in the DGGE gel. As can be seen by comparison of the melting map, the restriction

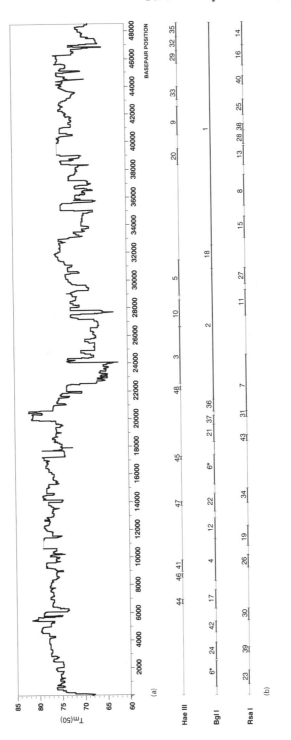

Fig. 4.2. (a) Melting map of the complete lambda genome (48 502 basepairs). The $T_m(50)$ of each basepair (see Chapter 1) is plotted against its position in the DNA sequence of the genome. (b) Below the melting map all Hae III, Rsa I and Bgl I restriction fragments larger than 500 bp are schematically corresponding to their position in the lambda genome. Numbers correspond to the fragments in the schematic 2D separation pattern shown in (c). 6* are Bgl I fragments co-migrating in the 2D gel.

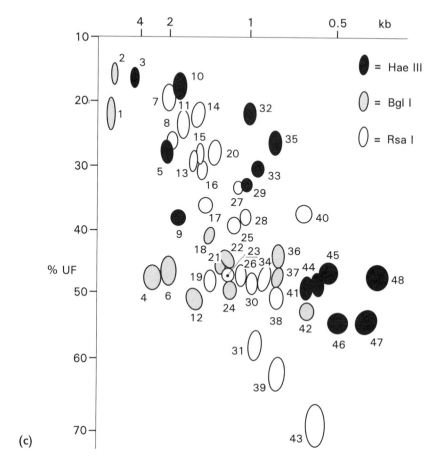

Fig. 4.2. (c) 2D gel separation pattern of a mixture of phage lambda fragments generated after separate digestion with the Hae III, Rsa I and Bgl I restriction enzymes. The 2D gel was composed of 6% PAA and contained a gradient of 10–75% UF; electrophoresis was performed at 60°C for 12 h at 200 V. Numbers correspond to the fragments schematically depicted in (a).

map and the separation patterns of the lambda DNA restriction fragments, all details of the separation pattern observed are determined by the number, length, structure and distribution of the melting domains within a given genome. These characteristics will determine the resolution obtained in a 2D separation pattern of any given DNA molecule.

A more complex example of 2D DNA typing of lower organisms is the analysis of mycoplasma and bacterial genomic DNA. The size of these genomes can vary from 500 kb up to 5000 kb as established by renaturation kinetics, PFGE mapping and two-dimensional DNA electrophoresis using double restriction enzyme digestion (see also Chapter 1). The number of restriction fragments obtained from these organisms can be close to the upper limit of what can be separated on a 18×15 cm standard 2D gel (in theory about 1000 spots. Yet, 2D DNA typing using DGGE has been successfully applied to obtain estimates of the total number of restriction fragments and hence of the genome size of the mycoplasma *Mycoplasma capricolum* (724 kb) and *Acholeplasma laidwaii* (1646 kb) and the bacterium *Hemophilus influenzae* (1833 kb) (Poddar and Maniloff, 1989).

In Fig. 4.3 a 2D separation pattern is shown of genomic DNA from *Bacillus amyloliquefaciens* digested with the restriction enzyme Hinf I. Using a 1D size separation in long (30 cm) 4–9% PAA gradient gels and a narrow denaturing gradient (25–55% UF) in the second dimension gel, it was possible to resolve about 800 restriction fragments in a spot pattern occupying almost the complete 2D gel. Based on these and comparable results obtained with other digests (Rsa I and Eco RI/Hind III) this number of fragments was calculated to correspond to a genome size of roughly 4000 kb.

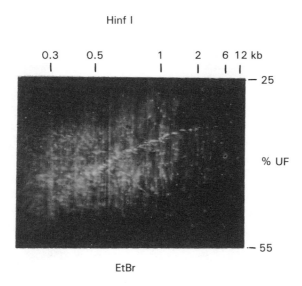

Fig. 4.3. Ethidium-bromide stained 2D gel separation pattern of fragments of the *Bacillus amyloliquefaciens* genome generated after digestion with Hinf I. The 2D gel was composed of 6% PAA and contained a gradient of 25–55% UF; electrophoresis was performed at 58°C for 12 h at 200 V.

4.2.2 Analysis of lower eukaryotes

The relatively small size of the viral and bacterial genome allows direct detection of all fragments, resulting from restriction enzyme digestion, on a 2D gel of proper dimensions with relatively insensitive ethidium bromide staining. Ethidium bromide staining of two-dimensional separations of restriction enzyme digests of genomic DNA isolated from lower eukaryotes such as yeast results in an unresolvable cloud with a few bright spots. (The latter are derived from repetitive elements present in the genome, either clustered or dispersed, which will generate several (nearly) identical restriction fragments.) Rather than resolving the entire genome, genome scans can be made of these organisms using hybridization analysis with repeat element probes. Since genomes of lower organisms are usually much more compact than those of higher organisms, the choice of probes detecting such elements is limited and strongly dependent on the organism. For the yeast *Saccharomyces cerevisiae* several transposable elements have been described such as the δ-element, the Σ element and other related transposable elements (for reviews, see Moroz–Williamson, 1983; Genbauffe *et al.*, 1984).

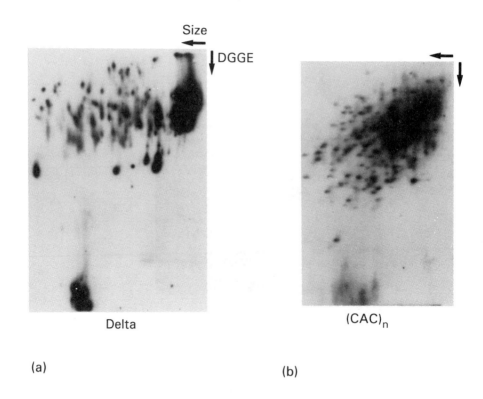

(a)   (b)

Fig. 4.4. (a) Hybridization analysis using a probe for the δ-transposon element of a 2D gel separation pattern of fragments of the *Saccharomyces cerevisiae* genome generated after digestion with Hae III. The 2D gel was composed of 6% PAA and contained a gradient of 25–55% UF; electrophoresis was performed at 58°C for 12 h at 200 V. (b) Hybridization analysis with microsatellite core probe $(CAC)_n$ of a 2D gel separation pattern of fragments of the *Aspergillus niger* genome generated after digestion with Hae III. The 2D gel was composed of 6% PAA and contained a gradient of 25–55% UF; electrophoresis was performed at 58°C for 12 h at 200 V.

An example of a hybridization pattern of *Saccharomyces cerevisiae* DNA analysed by 2D DNA typing using a δ-specific probe is shown in Fig. 4.4a. About 120 spots can be observed in this Hae III digest and comparable numbers have been observed with other enzymes. This indicates the existence of a similar number of loci with this repeat element in the yeast genome. Also with some micro- and minisatellite core probes spot patterns were obtained, indicating the widespread occurrence of this type of satellite motif in the yeast genome. For example, with microsatellite probe $(CAC)_n$ clear spot patterns consisting of about 200 spots were observed (results not shown).

Analogous to yeasts, other lower eukaryotes can be analysed by 2D DNA typing. In Fig. 4.4b a hybridization pattern is shown obtained with the microsatellite core probe $(CAC)_n$ of a Hae III digest of genomic DNA from *Aspergillus niger* genomic DNA. As in yeast, about 200 spots were observed with this probe. Comparable spot patterns were obtained for *Aspergillus* with minisatellite core probe 33.6 and the microsatellite $(GT)_n$ (results not shown). These results indicate that tandemly repetitive elements occur in many different lower eukaryotes and can be used for genome scanning studies.

4.3   GENETIC LINKAGE ANALYSIS

4.3.1   Introductory remarks

Genetic linkage analysis has been shown to be a powerful tool in the elucidation of the molecular basis of human genetic diseases (Botstein *et al.*, 1980; White *et al.*, 1985; see also Chapter 2). The approach involves the analysis of many polymorphic loci spread over the genome of interest. These loci serve as marker loci to study co-segregation with other marker loci, to generate a marker map, and with particular phenotypes to localize genetic traits. The number of loci to be analysed depends on the size of the genome, in that usually one marker locus every 5–10 million basepairs has to be present. This number is necessary in order to reach statistically significant levels of certainty in concluding linkage between the marker and the genetic trait locus. For the human genome, with an estimated size of 3×10^9 bp, the number of markers to be analysed in order to find linkage within 5×10^6 bp would be roughly 600, provided each marker is able to generate informative variants in the pedigree which is under scrutiny. For the human and mouse genomes many highly polymorphic $(GT)_n$ microsatellite markers are now available (Weissenbach *et al.*, 1992; Dietrich *et al.*, 1992), but the amount of work involved in screening the corresponding loci is still considerable (Hearne *et al.*, 1992).

In view of the limited resolving power of one-dimensional gel electrophoretic analysis which is the basis of all currently available scanning systems, all markers have to be analysed one by one. Depending on the nature of the marker systems analysed, some compression of effort can be achieved by analysing more samples per lane. Indeed, it has been reported that by taking a so-called multiplex approach up to 16 microsatellite marker loci can be analysed per lane (Todd,

1992). Nevertheless, the number of pipetting steps involved in the processing of large numbers of individual markers is very high. Although it has been claimed that extensive automation will eventually solve all these problems (Todd, 1992) it is still an open question as to how successful large centralized semi-automated laboratories will be in solving the genetics of complex traits (see below). Two-dimensional DNA typing could be an alternative way of gene mapping, which may be more within reach of the common laboratory and still competitive with the factory science that is widely advocated as the way to go in the genome projects.

In Table 4.1 a comparison is made between locus-specific microsatellite typing and 2D DNA typing with micro- and minisatellite core probes highlighting the disadvantages and advantages of the latter system (see also Table 2.2). The main advantage of 2D DNA typing is obviously the high information density per gel; several core probes detect 300–400 spots corresponding to a similar number of alleles. Simply by rehybridizing each filter about 20 times with different non-overlapping core probes a theoretical map resolution of about 1 million basepairs can be reached. In practice, the effective resolution will be lower because not all spots detected represent alleles from polymorphic loci.

Table 4.1. Evaluation of 2D DNA typing[a]

Advantages of 2D DNA typing:
(1)   high information density per gel;
(2)   very few pipetting steps;
(3)   no prior genetic map required.

Disadvantages of 2D DNA typing:
(1)   no immediate information on map location of markers;
(2)   technically demanding.

[a]Based on comparison of 2D DNA typing with multiplex $(GT)_n$ VNTR microsatellite single-locus typing.

Table 4.2 lists the numbers of spots, percentages of variant spots and spot overlap for several core probes tested in 2D DNA typing of humans. The overlap in spot positions among these core probes appears to be low, on average. In view of the high resolving power of the gel system it is unlikely that this low overlap is due to co-migration of fragments. Rather, cross-hybridization of one micro- or minisatellite to two or more core probes will occur. For some particular probes extreme overlap was found with minisatellite 33.6. For example, HBV-1 almost totally overlaps with 33.6, while HBV-2 and $(TCC)_n$ also show some overlap with this probe (results not shown). From Table 4.2 it might appear that the percentage of variant spots (alleles from polymorphic loci) is rather low. However, it should be realized that the percentages listed there are all derived from comparisons between two randomly chosen individuals. For locus-specific probes heterozygosity levels are usually derived from 50–150 unrelated individuals. Results obtained from eight unrelated individuals with minisatellite probes 33.6 and 33.15 already indicate more than 80% variant spots. Such 2D patterns represent highly informative genetic marker systems (see also 4.3.3).

TABLE 4.2. Core probe characteristics in 2-D DNA typing of humans[a]

(A) *Numbers of spots and percentages of spot variants*

Probe	Spots	Spot variants (%)
33.6	456	20
33.15	482	22
INS	430	12
HBV-1	345	34
HBV-2	140	ND[b]
HBV-5	350	ND[b]
ENHA	200	ND[b]
M13	122	ND[b]
HBV-3	75	ND[b]
CAC	607	16
TCC	175	ND[b]
AGC	109	13
TELO	82	20
Total	3573	

(B) *Overlap among core probes[c]*

	33.6	33.15	CAC	HBV1	Number of spots
33.6		7%	6%	60%	456
33.15	9%		12%		482
CAC	5%	9%			607
HBV1	80%				345

[a]Based on the analysis of at least two unrelated individuals.
[b]ND = not determined.
[c]Based on the analysis of six unrelated individuals.

The main disadvantage of 2D DNA typing is the lack of immediate information as to the identity of the spots detected by these core probes. The situation can be understood as follows. Each spot in a 2D DNA typing pattern represents a single or double copy of a specific allele, or represents two or more unrelated alleles which fortuitously co-migrate in the 2D gel. In comparison, locus-specific microsatellite typing reveals information about all specific alleles of a certain locus. Linkage analysis of 2D DNA typing data requires that single unknown alleles can be specified. Unfortunately, the presence of a spot does not tell whether heterozygosity or homozygosity exists for the presence of the corresponding allele. Nevertheless, 2D DNA typing data contain genetic information. For example, when two parents have a spot at a specific position in the 2D gel which is lacking in a child, it can be inferred that the parents share one allele. Compared with a fully informative marker less information is available, but enough to be useful for linkage analysis. In this respect we have shown that 2D DNA typing data can be introduced directly into linkage calculations and demonstrated the use of 2D DNA typing data by showing linkage of a particular spot variant with a ge- netic trait in cattle (te Meerman *et al.*, 1993; see also below). In general, the results indicate that the power to detect linkage by 2D DNA typing becomes equivalent to that of an analysis using locus-specific markers by increasing the sample size by a factor of two (te Meerman *et al.*, 1993).

To bring 2D DNA typing to the same level of informativity as single-locus typing, two possibilities can be envisaged. First, it is possible to find complementary alleles (from the same locus) by studying co-segregation of 2D DNA typing spots with themselves. Such complementary alleles will segregate in repulsion in a pedigree. This approach to detect complementary alleles is greatly facilitated by the observation that alleles from the same locus, but of different length, migrate to one isotherm in the denaturing gradient (second dimension) gel, except when one or more VNTR repeat units constituting the allele contain a recognition site of the restriction enzyme used to digest the genomic DNA (Uitterlinden and Vijg, 1991a; Hovig *et al.*, 1993; see below). Digestion of genomic DNA with a different enzyme can solve this problem. In addition, the genetic information of 2D DNA typing data can be increased by identifying closely linked spots, which will often co-segregate in a pedigree. Several closely linked spots can then be treated as a highly informative haplotype.

Second, spots of interest can be directly isolated from the 2D DNA typing pattern and developed into locus-specific markers (de Leeuw *et al.*, in preparation). Fig. 4.5 illustrates one approach that can be followed to isolate the marker-containing DNA fragments indicated by the spots. Fig. 4.5a shows the use of duplicate gels to excise the gel pieces corresponding to the spots of interest indicated by the hybridization patterns obtained with the master gel. After elution from the gel the genomic DNA fragment is subsequently ligated into a plasmid and subjected to a PCR procedure to amplify specifically the flanking sequences of the marker locus. One of the primers used is the core sequence itself; the others are M13 priming sites on the plasmid used (Fig. 4.5b). Fig. 4.5c illustrates the procedure by showing the duplicate gel with the genomic DNA digest and the marker mixture (left), a 33.15 hybridization pattern obtained from the master gel (middle) and the marker hybridization pattern of the same master gel (right). The markers are used to localize the spots of interest on the duplicate gel. Fig. 4.5d shows the results obtained with the same master gel after hybridization to PCR-amplified flanking sequences of a particular marker isolated as described above. The spot that was isolated is indicated in Fig. 4.5c by an arrow (middle). In Fig. 4.5d the same spot can be observed. (Note that these flanking sequences are specific to only one locus rather than hybridizing to the whole set of 33.15-specific minisatellites.) On top of Fig. 4.5d the one-dimensional hybridization pattern obtained with this probe is shown. Two bands can be observed. One is at the same size position as the spot in the two-dimensional pattern, while the small second band (the complementary allele) was not included in the second dimension gel. In general, the use of such markers as probes into 2D DNA typing blots will subsequently reveal complementary alleles. At present, routinely applicable protocols for spot isolation from 2D gels are being developed.

Before showing some examples of the use of 2D DNA typing in pedigree analysis in humans and animals, we will first discuss in some detail the electrophoretic behaviour of GC-rich minisatellite fragments in denaturing gradient gels.

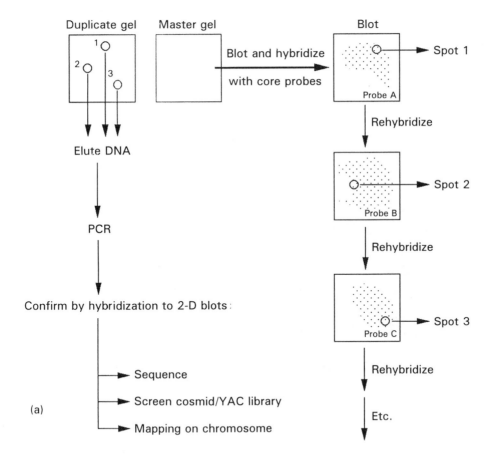

Fig. 4.5. (a) Schematic depiction of the strategy that can be followed to develop spots of interest in 2D DNA typing hybridization patterns into locus-specific markers.

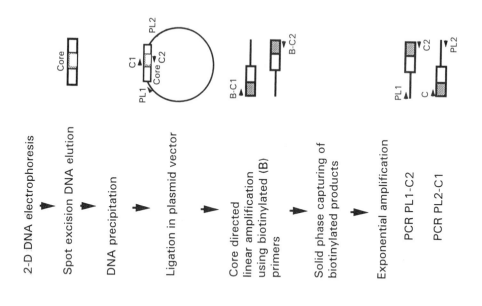

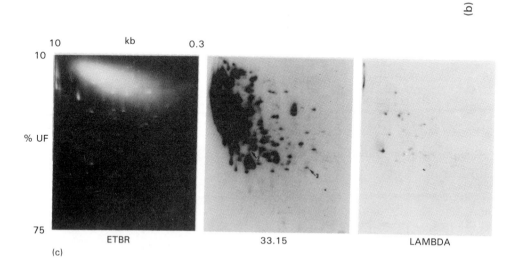

Fig. 4.5. (b) Outline of a PCR-based procedure to obtain micro- and minisatellite flanking sequences directly from excised 2D DNA typing spots. (c) Spot isolation procedure. The spot of interest (3, arrow) is indicated in the 33.15 hybridization pattern obtained with the master gel (middle). On the basis of the marker pattern obtained from the master gel by hybridization with lambda probe (right) and visible in the duplicate gel after ethidium bromide staining (left) the spots of interest are excised from the gel.

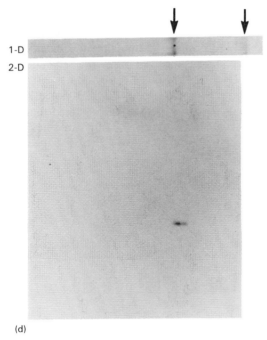

(d)

Fig. 4.5. *(continued)* (d) Hybridization analysis with the amplification product obtained from the spot of interest (3) as probe of Hae III digested genomic DNA of the same individual which was used to isolate the locus specific probe, after one-dimensional size separation in a PAA gel (1D) and as present in the 2D master gel (2D). The two arrows indicate the two alleles present in this individual for this locus. In this particular case the (very small) complementary allele was not included in the second dimension separation.

4.3.2   Locus-specific melting behaviour of minisatellite alleles

Each of the spots detected in a 2D DNA typing pattern of an individual represents one of the two alleles of a micro- or minisatellite locus detected by the core probe used. Deviations from this rule can occur, when (a) one spot represents two identical 'alleles' (not all mini- or microsatellite loci are heterozygous) or (b) the mini- or microsatellite units have a recognition site for the restriction enzyme used, resulting in several spots per allele. Alleles of highly polymorphic VNTR loci may greatly differ in size, but display locus-specific electrophoretic behaviour in denaturing gradient gels. This is illustrated by denaturing gradient gel electrophoretic analysis of several minisatellite alleles. The melting map of an allele of the D7S22 locus (detected by pλG3) containing 14 repeat units (Fig. 4.6) shows that the minisatellite region represents a high melting domain (HMD), whereas the adjacent sequences form low melting domains (LMD). Of the LMDs the 5′ one, between the Alu I repeat and the simple sequence region, represents the lowest melting domain. This domain is included in an Hae III restriction fragment encompassing the minisatellite region.

Therefore, it can be assumed that this domain will determine the position of the Hae III restriction fragment after denaturing gradient gel electrophoresis (DGGE), irrespective of the size of the allele. (Hinf I sites fall outside the sequence shown in Fig. 4.6, so it is unknown whether Hinf I fragments share the 5′ lowest melting domain with Hae III fragments.) In addition, several other G+C-rich minisatellite sequences (such as the one in the myoglobin region; Fig. 4.7) were found to represent HMDs. It can therefore be predicted that in general G+C-rich minisatellite-containing Hae III and/or Hinf I fragments having the same flanking sequences because they are from the same locus will migrate to the same position in a denaturing gradient gel.

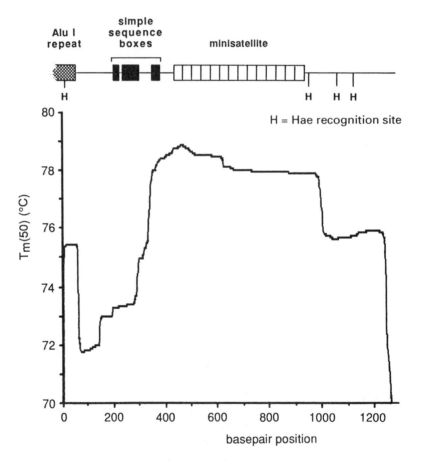

Fig. 4.6. Calculated melting map of the D7S22 VNTR locus. The sequence was taken from Wong *et al.* (1986) with the minisatellite consisting of fourteen 37 bp repeat units (the first five repeat units as published and the following nine according to the consensus sequence with Y=T and R=A) (from Uitterlinden and Vijg, 1991).

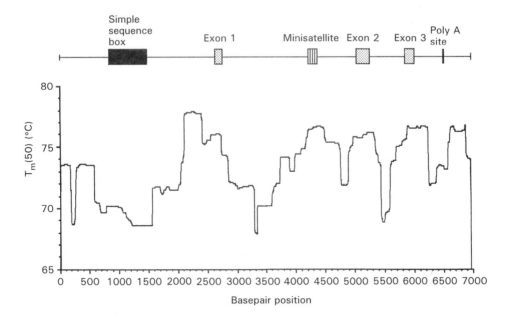

Fig. 4.7. Calculated melting map of the myoglobin minisatellite locus. The sequence data
were obtained from the EMBL library.

These theoretical predictions were experimentally tested by analysing Hae III
and Hinf I digests of genomic DNAs from unrelated human individuals with
respect to the electrophoretic behaviour of minisatellite-containing fragments in
denaturing gradient gels (Uitterlinden and Vijg, 1991). For that purpose genomic
digests were separated by DGGE and electrotransferred to a nylon membrane,
which was subsequently hybridized with inserts from locus-specific VNTR
probes.

In Fig. 4.8 results are shown for the Hinf I digests hybridized with pλG3 and
pYNH24 detecting the loci D7S22 and D2S44, respectively (Uitterlinden and
Vijg, 1991). For both these polymorphic loci, alleles of different sizes were
observed after Southern analysis of the individuals tested. After DGGE, however,
only a single hybridizing band could be observed at 38% and 39% denaturant
concentration for pλG3 and pYNH24, respectively. The position of this band
was identical among the individuals for pYNH24, but varied slightly for pλG3
within a 3% denaturant concentration range.

Fig. 4.9 shows the results obtained for the Hae III digests of 33 individuals
after hybridization with the pλG3 probe. The presence of more than two bands
per lane after Southern blot analysis indicates the presence of Hae III recognition
sites within the minisatellite region of this locus. For pλG3 an intense band was
observed in the Hae III DGGE gel at 38% denaturant concentration which we
infer to be derived from Hae III fragments containing the 5′ LMD plus the

minisatellite sequence or part thereof. The allelic fragments containing the 5′ domain (which range in size from 2.5 kb to about 14 kb) co-migrate except for 9 individuals having the short common allele (see Wong *et al.*, 1986 and 1987), visible as a very sharp band at 42% denaturant concentration (note the homozygous individual 31 as indicated by the one sharp band in the DGGE analysis). This aberrant DGGE behaviour might be due to a particular sequence difference in the 5′ LMD which all small common alleles have in common in contrast to the 5′ LMD in larger alleles. The minor variations in gradient position of the larger alleles containing the 5′ LMD do not correlate with their length and are therefore most likely due to sequence variations in the 5′ lowest melting domain. The variations in gradient position found after Hae III digestion were also observed with Hinf I digests, albeit within a more narrow range (see Fig. 4.8).

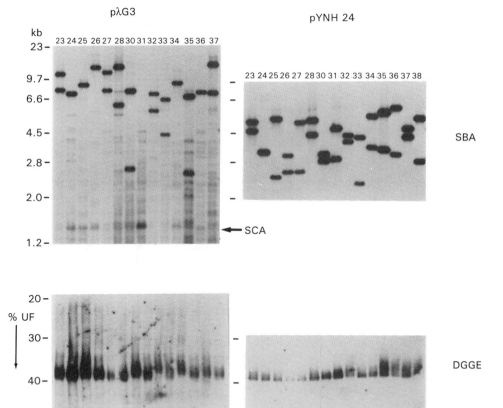

Fig. 4.8. Hybridization analysis of the D7S22 and D2S44 loci, detected by pλG3 and pYNH24, respectively, in genomic DNA from 14 unrelated individuals. Genomic DNA was digested with HinfI and separated by neutral agarose gel electrophoresis (Southern blot analysis; SBA) and by parallel denaturing gradient gel electrophoresis (DGGE). SCA = Short common allele (from Uitterlinden and Vijg, 1991).

Of the 33 individuals, 12 (36%) had internal Hae III sites in the VNTR sequences of one or both pλG3 alleles, as indicated by the presence of more than two bands in the Southern blot analysis (SBA) (see Fig. 4.9a). The presence of occasional Hae III sites in the minisatellite region of this locus has been noticed before by Wong *et al.* (1987). This was also reflected by the presence of one or two bands in the DGGE gel at about 50% denaturant concentration (indicated by a–f). These fragments contain only the G+C-rich minisatellite sequences and no adjacent sequences. Most likely, the presence of internal melting domains in the minisatellite region results in their migration to positions at higher denaturant concentrations in the denaturing gradient gel. The slight variation in the position of these bands in the denaturing gradient is size independent and suggests differences in sequence composition of minisatellite repeat units. We could discern six variants a–f (see Fig. 4.9a). The low frequency at which these variants occur, together with the low incidence of internal Hae III sites in the minisatellite region, indicates that the repeat units of VNTR alleles from this locus have a high level of sequence homogeneity.

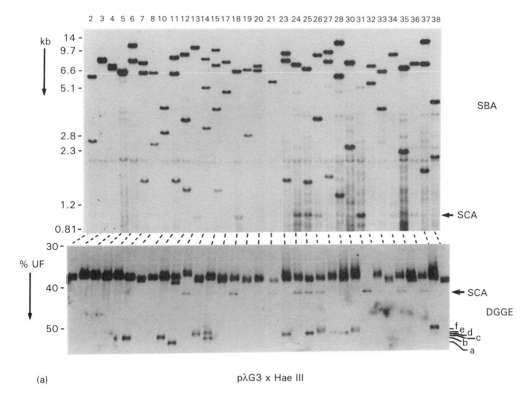

(a)   pλG3 x Hae III

Fig. 4.9. (a) Hybridization analysis of the D7S22 VNTR locus as detected by probe pLamb-daG3 in genomic DNA from 33 unrelated individuals. Genomic DNA was digested with HaeIII and separated by neutral agarose gel electrophoresis (SBA) and by parallel denaturing gradient gel electrophoresis (DGGE). SCA = Short common allele; a–f indicates different internal fragments (from Uitterlinden and Vijg, 1991).

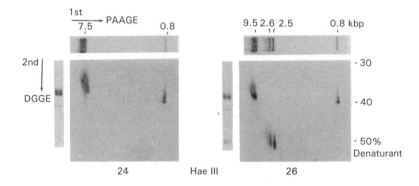

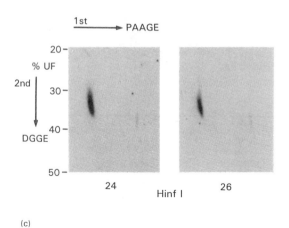

(b)

(c)

Fig. 4.9. (b) Two-dimensional gel analysis of the D7S22 locus in genomic DNA from two unrelated individuals after Hae III digestion. Also the one-dimensional size separations in neutral polyacrylamide gels and one-dimensional parallel denaturing gradient separations are shown, which were run in separate experiments. (c) Two-dimensional gel analysis of the D7S22 locus in the same individuals as in (a), but this time after Hinf I digestion (from Uitterlinden and Vijg, 1991).

In order to obtain information on the origin of restriction fragments observed in parallel DGGE we analysed genomic digests by two-dimensional electrophoresis. Figs 4.9b and 4.9c show two-dimensional gel analyses of the D7S22 locus in Hae III and Hinf I digested genomic DNA from two unrelated individuals. As already demonstrated by one-dimensional parallel DGGE (Fig.

4.9a) the short common allele (SCA; 0.8 kb) migrates to a slightly lower gradient position as compared with the Hae III fragments derived from the much larger alleles (7.5 and 9.5 kb). The two extra Hae III fragments of 2.5 and 2.6 kb in individual 26 migrate further into the gradient. As already suggested by the data shown in Fig. 4.8a, they most likely represent restriction fragments which are generated as a consequence of two internal Hae III sites in the minisatellite sequences of the large allele in this individual (note that the two fragments appear as a single band of about 2.5 kb in agarose Southern blot analysis). This was confirmed by using Hinf I to digest the same genomic DNA samples, whereby only two spots were obtained under identical hybridization conditions, corresponding to the large and small allele, and no internal fragments were observed (Fig. 4.9c).

Similar observations on the isothermal DGGE behaviour of alleles of a VNTR locus have been made for a minisatellite in the retinoblastoma gene (Hovig *et al.*, 1993).

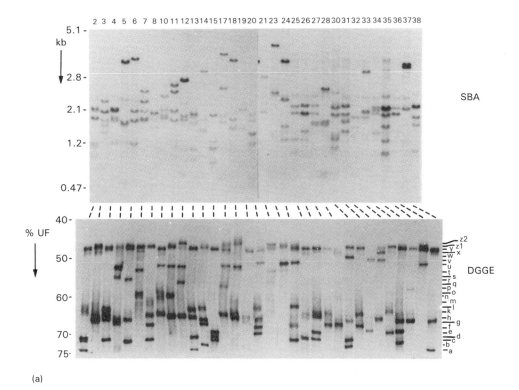

(a)

Fig. 4.10. (a) Hybridization analysis of the D7S21 VNTR locus as detected by probe pMS31 in genomic DNA from the same 33 individuals as in Fig. 4.8a after Hae III digestion. Separation was in neutral agarose gels (SBA) and in denaturing gradient gels (DGGE). a–w indicate different internal fragments; x–z2 are fragments most likely containing locus-specific LMDs outside the minisatellite array of this locus.

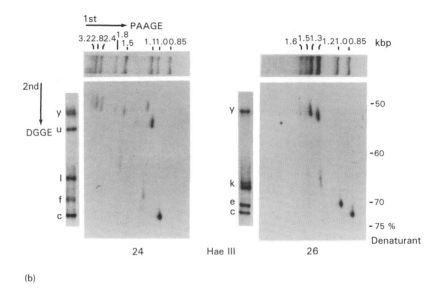

(b)

Fig. 4.10. (b) Two-dimensional gel analysis of the D7S21 VNTR locus in two unrelated individuals after Hae III digestion. As in Fig. 4.8, the one-dimensional separation patterns are also shown.

Another example of more than two spots in a 2D pattern being derived from one VNTR locus is shown in Fig. 4.10 for the D7S21 locus detected by probe pMS31 (Wong *et al.*, 1987). In this case up to 10 spots (individual 24) could be observed for one locus after Hae III digestion. As can be seen in Fig. 4.10a internal Hae III recognition sites occur very frequently for this locus. Two groups of fragments can be distinguished. The upper fragments around 50% UF presumably contain flanking sequences and, hence, show isothermal behaviour. The lower fragments of between 55% and 75% UF may represent the GC-rich internal minisatellite fragments. It is obvious that the frequent occurrence of such loci greatly diminishes the information content of 2D DNA typing patterns. However, thus far we have no indication that the phenomenon is widespread among the rather small mini- and microsatellite loci that fall within the resolution range of 2D DNA typing.

The results provided above indicate that alleles of different sizes of a VNTR locus of the G+C-rich set of minisatellite sequences can display very similar electrophoretic behaviour in denaturing gradient gels. This is due to the fact that alleles of the same locus share a relatively monomorphic low melting domain outside the highly polymorphic minisatellite. The position in the denaturing gradient is determined by this low melting domain; the minisatellite sequence effectively functions as a clamp, that is, a high melting domain that stabilizes the molecule (Lerman *et al.*, 1984; Myers *et al.*, 1985b). The choice of restriction enzymes to digest genomic DNA is of importance: we found Hae III to give rise to more complex DGGE separation patterns which is most likely due to the presence of internal recognition sites in the minisatellite at this locus.

The isothermal electrophoretic behaviour in DGGE gels of VNTR alleles facilitates the identification of alleles in 2D DNA typing by genetic means. However, to benefit optimally from this phenomenon it is necessary to use a restriction enzyme with no recognition sites in the VNTR alleles detected by the core probes. The above shows that Hae III is not always the right choice but more extensive studies will show how frequently this is the case among the micro- and/or minisatellite alleles detected by a core probe.

The isothermal behaviour of GC-rich micro- and minisatellite alleles in denaturing gradient gels is important for grouping all spots detected by a given core probe in combination with a given restriction enzyme in allelic pairs. Ultimately this will result in a genome map (see below). However, the phenomenon also explains why the GC-rich repeat families are suitable for obtaining high-resolution spot patterns covering the entire gel. Indeed, ethidium bromide staining of a typical (Hae III or Hinf I) 2D separation pattern shows that most of the fragments are clustered rather high in the denaturing gradient (Fig. 4.11, left). This indicates the presence of a low melting domain in these fragments, present for example in the Alu repeat unit present in most genomic restriction fragments of this size (Lerman *et al.*, 1986). Indeed, the pattern is indistinguishable from the hybridization pattern with Alu as probe (Vijg and Uitterlinden, 1991).

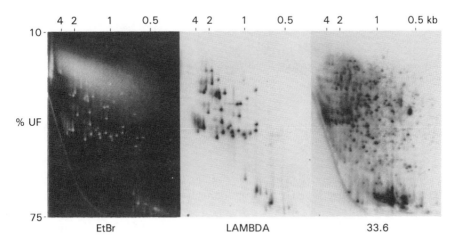

Fig. 4.11. Two-dimensional DNA typing patterns obtained with human genomic DNA mixed with the lambda fragments also shown in Fig. 4.2 (see 4.2.1). On the left the ethidium bromide stained pattern is shown with the cloud of human DNA and the clear spots of the markers. The middle shows the markers again, but this time as an autoradiograph after hybridization of the gel pattern with lambda fragments as probe. The right shows the pattern obtained after hybridizing the gel pattern with minisatellite core probe 33.6.

Fig. 4.12 shows the hybridization pattern obtained with core probe (GT)$_n$ as an example of a suboptimal separation. Most of the fragments are clearly clustered high in the denaturing gradient gel, while near the edges some spots are resolved. Obviously, in human DNA this probe detects a large number of fragments. In lower eukaryotes, such as *Aspergillus*, only around 200 spots were obtained, well resolved but almost all present in the upper regions of the denaturing gradient gel (results not shown).

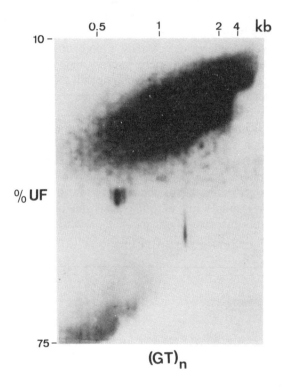

Fig. 4.12. Two-dimensional DNA typing pattern of human genomic DNA after hybridization with microsatellite core probe (GT)$_n$.

In summary, it appears that in the fragments detected by GC-rich probes the repeat units act as the highest melting domains and therefore effectively as a GC-clamp. By contrast, AT-rich fragments will act as early melting domains, which explains the clustering of those fragments relatively high in the gel.

4.3.3   Anatomy of a 2D DNA typing gel

Based on the data shown in the previous sections we will now describe in some more detail the quality and quantity of the genetic information as contained in a 2D DNA typing gel and how to interpret spot patterns and variations therein. In general a micro- and minisatellite core probe detects monomorphic micro-

and minisatellite regions and polymorphic regions which correspond to the VNTR type of genetic markers. This will result in a spot pattern corresponding to a certain number of genetic loci.

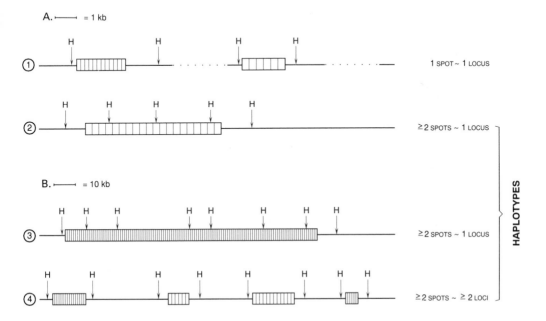

Fig. 4.13. Schematic representation of the correspondence between spots in a 2D DNA typing pattern and the alleles at a VNTR locus (H = Hae III recognition site); A and B denote differences in scale. (1) three situations can be discerned: 1 spot = 1 allele (heterozygosity); 1 spot = 2 alleles (homozygosity); co-migration of two VNTR alleles from different loci, sharing the core probe sequence. (2) ≥2 spots = 1 allele, owing to the presence of Hae III recognition sites in some of the repeat units of the micro-/minisatellite region. (3) As for (2), but now in a very large 'midisatellite' allele resulting in several spots belonging to the same haplotype. (4) closely linked VNTR loci sharing the same core probe sequence.

Ideally each of the spots corresponds to one allele of a VNTR locus which can be used as an independent genetic marker. However, based on the analysis of individual VNTR loci (4.3.2) several deviations from this ideal situation have been found. Fig. 4.13 schematically depicts the correspondences theoretically possible between micro- or minisatellite loci and spots observed in a 2D DNA typing pattern generated by a core probe. The different situations are discussed below.

(1)   The simplest situation is one 2D spot corresponding to one allele of a VNTR locus. The incidence of a particular 2D spot depends on the frequency of the corresponding allele of the VNTR locus in the population. Highly polymorphic loci with multiple low-frequency alleles will give rise to spots with similar

characteristics while apparently monomorphic spots can turn out to originate from a VNTR with one frequently occuring allele and with few alleles, which have a very low incidence in the population. If the individual is homozygous for a VNTR locus one spot will correspond to two identical alleles, a situation which cannot readily be distinguished from fortuitous co-migration of two (or more) alleles from different loci (due to identical size and melting temperature). In view of the high resolving power of 2D DNA typing, the exquisite sequence sensitivity of DGGE, and the limited number of loci detected by one core probe (only a few hundred), the latter situation is unlikely to occur.

(2)   The presence of recognition sites of the restriction enzyme used, in two or more of the repeat units of the VNTR locus detected by a core probe, can account for the observation that more than one spot corresponds to one allele (see 4.3.2). In pedigree analysis such fragments will therefore segregate as a haplotype.

(3)   Such haplotypes can include the so-called 'midisatellites': minisatellite types of regions which can be up to several hundred kilobases in length (Buroker et al., 1987; Knowlton et al., 1989).

(4)   Because of very tight local clustering of some VNTR loci (Royle et al., 1988; Giannakudis et al., 1992; our own unpublished observations) two or more spots can derive from alleles of closely linked VNTR loci and segregate as a haplotype much like the situation described under (3).

These different types of correspondences between 2D DNA typing spots and alleles of loci scored lead to different types of spot variants (or allelic states of 2D spots) which can be observed in 2D DNA typing analysis of pedigrees. These are schematically depicted in Fig. 4.14. It should be noted that currently several of the allelic states of 2D spots cannot be readily distinguished owing to the relatively low sensitivity of the 2D DNA typing procedure to detect spot intensity differences of a factor 2.

Estimated fractions of the variants observed in the analysis of a limited number of individuals and the corresponding pedigrees are also provided (Fig. 4.14). So far we have found similar numbers for both cattle and human DNA. When two unrelated individuals are compared, on average about 20–30% of the total number of spots is variant for the multilocus core probes, such as 33.6, 33.15 and $(CAC)_n$ (see also Tables 4.2 and 4.4). We have observed that the total number of spot variants detected by a given core probe is increasing when more individuals are analysed. Our estimate now is that at least 80% of the spots detected are in fact variant in the population and hence derived from polymorphic loci. A plateau valueof the percentage of spot variants is slowly reached at 8 individuals (Mullaart et al., in preparation). From pedigree analysis it appears that about 70% of the spot variants are transmitted in equal proportions as either paternal or maternal spot variants (Uitterlinden et al., 1989a; Mullaart et al., in preparation). This number is higher than expected when each spot represents an independently assorting locus, which is most likely due to the occurrence of haplotypes as described above. In addition, spots representing identical alleles (homozygous state) from an otherwise polymorphic locus, will contribute since such a spot will always be transmitted. Mutant spots have until now been rarely observed (see also below, 4.3.4.1).

	2-D PATTERN[1]			ALLELIC STATE[2]			FRACTION[3]
	MOTHER	FATHER	CHILD	MOTHER	FATHER	CHILD	
	"CONSTANT" SPOTS						
1	●	●	●	HOM	HOM	HOM	
2	•	●	●	HET	HOM	HOM	
3	●	•	●	HOM	HET	HOM	
4	•	●	•	HET	HOM	HET	≤ 20%
5	●	•	•	HOM	HET	HET	
6	•	•	●	HET	HET	HOM	
7	•	•	•	HET	HET	HET	
	SPOT VARIANTS						
8	•	○	•	HET	?	HET	35%
9	●	○	•	HOM	?	HET	MATERNAL
10	○	•	•	?	HET	HET	35%
11	○	●	•	?	HOM	HET	PATERNAL
12	•	○	○	HET	?	?	10%
13		•	○	?	HET	?	NON-TRANSMITTED
14	•	•	○	HET	HET	HET	≤ 1 %
15	○	○	•	?	?	MUT	≤ 0.001 %

1   ●   =   Two copies of micro-/minisatellite containing restriction fragment
    •   =   One copy of micro-/minisatellite containing restriction fragment
    ○   =   Absence of spot

2   HOM   =   Homozygous
    HET   =   Heterozygous
    MUT   =   Mutation
    ?   =   Homozygous or heterozygous for other allele(s)

3   Expressed as estimated fraction of total number of spots observed in a pattern of one individual, measured in 2 CEPH pedigrees

Fig. 4.14. Schematic overview of allelic states of 2D spot variants as observed in pedigree analysis. See text for details.

On the basis of the above we can now look in a more knowledgable way at a 2D spot pattern. Fig. 4.15 shows a 2D DNA typing pattern obtained with multilocus core probe 33.15. In this pattern the particular features, as described in the previous sections, have been indicated and are discussed in detail below.

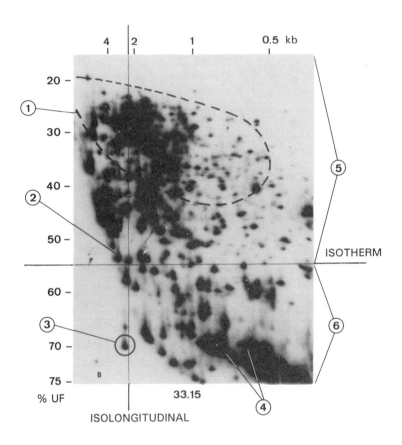

Fig. 4.15. Features of a 2D DNA typing spot pattern. Shown is the 33.15 hybridization pattern of a human Hae III digested genomic DNA sample. The different numbered features are discussed in the text.

(1)   Left of the 2 kb isolongitudinal we currently achieve suboptimal resolution: spots are relatively big, blurry and densely spaced and usually high background is observed. This feature can be explained by the large fragment size, preventing good separation in PAA gels, and by insufficient electrotransfer of these fragments to the hybridization membrane. From Southern blot analysis in agarose gels it appears that this area contains up to 20 very large VNTR alleles which are derived from very polymorphic loci. This limits the informativity of this area in 2D DNA typing, which is not a severe drawback in view of the small number of

fragments of this size. Also, above the 25% UF isotherm (in the upper 3 cm of the gel) a so-called 'electrophoresis gap' is frequently observed in which few spots are detected. The area arises due to local disturbance of the gel matrix after 6–8 h electrophoresis under standard conditions (see Chapter 3). In the upper left part of the pattern an area of relatively high background is observed which corresponds to the area where most genomic Hae III restriction fragments are present (indicated by the dashed line in Fig. 4.15). Their presence leads to more aspecific background hybridization than in other parts of the gel which are mostly devoid of restriction fragments (see also Fig. 4.1).

(2)   A diagonal of fragments can be observed on the very left bottom part of the 2D pattern. This corresponds to fragments which have not yet reached their melting point and still migrate according to size.

(3)   Occasionally we have observed (polymorphic) fragments which move faster than the diagonal of fragments moving according to size (see (3)). Currently we do not know the nature of this kind of fragment.

(4)   For some core probes we have observed additional diagonals in the right bottom part of the pattern. These are likely to correspond to (groups of) fragments which have already undergone complete melting and now migrate as single-stranded (ss) fragments more slowly than the double-stranded fragments. (The slower migration of ssDNA than the corresponding dsDNA can also be observed in perpendicular DGGE gels; see Chapter 1.) These spots are likely to be derived from so-called 'internal' fragments of particular VNTR loci (see 4.3.2 and below). They have only one melting domain (the melting temperature of which is shared by different fragments on the diagonal) and do not therefore focus at a particular place in the gradient but rather become directly single stranded on melting and continue migration through the gel but at a lower speed. The plus (G-rich)- and minus (C-rich)-strand of the minisatellite containing fragments might have different mobilities in this respect and, hence, result in (at least) two diagonals.

(5)   Above the 55% UF isotherm most if not all of the spots will carry a sequence flanking the micro- or minisatellite region. The melting domain of this flanking sequence will have a lower melting temperature and hence acts as the first melting domain to determine the position of the restriction fragment. If no 'internal' recognition sites for the restriction enzyme used are present in the alleles of the micro- or minisatellite region, two spots will be generated for one VNTR locus corresponding to the two alleles in the diploid genome (see 4.3.2).

(6)   Below the 55% UF isotherm many of the spots observed are likely to originate from the so-called 'internal' fragments generated from VNTR loci containing Hae III recognition sites in two or more of the repeat units of the allele (see 4.3.2). Some of these will have more than one melting domain and will therefore focus above the diagonal of completely melted fragments. These spots can be polymorphic (and as such can be used for genetic studies) but they will have one or more corresponding fragments above the 50% UF isotherm (which carry a flanking sequence acting as a first melting domain) and together segregate as a haplotype in genetic studies.

4.3.4   Examples

4.3.4.1 Humans
The applicability of the 2D DNA typing method in genetic studies on humans can be illustrated by the large number of transmitted variant spots that can be simultaneously followed in a two generation human pedigree of three members (Fig. 4.16; Uitterlinden *et al.*, 1989a). Close inspection and comparison of individual spot patterns of the two parents obtained with probe 33.6 revealed a total number of 569 spots for the father, 607 for the mother and 625 for the son. Between the two parents 150 spot variants were observed, 105 (=70%) of which were transmitted to the son (52 of maternal and 53 of paternal origin). Details of the separation pattern are shown in Fig. 4.16b. Using probe 33.6 we detected a fragment in the son which was not present in the mother or the father (Fig. 4.16c). This fragment is most likely a germ line mutation. Furthermore we observed four fragments, common to the parents, which could not be detected in the son (two examples are shown in Fig. 4.16c). Presumably these represent alleles of a VNTR locus shared by the parents while the other allele has been transmitted to the son. This indicates heterozygosity for this particular VNTR locus in the parents.

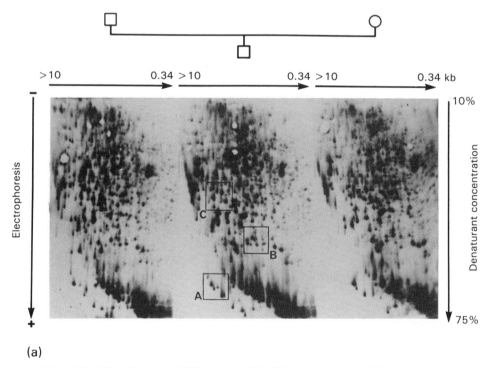

(a)

Fig. 4.16. (a) Two-dimensional DNA typing of Hae III digested genomic DNA from three members of a human pedigree, using probe 33.6 (from Uitterlinden *et al.*, 1989a).

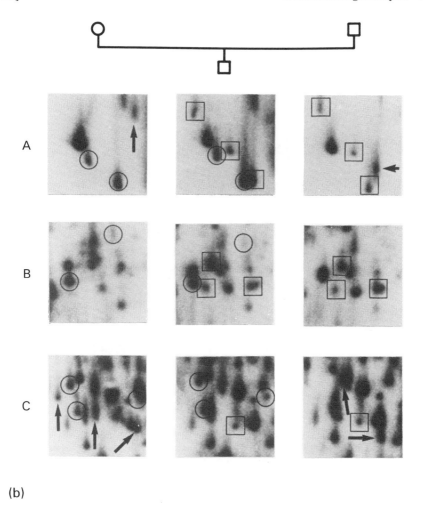

(b)

Fig. 4.16. (b) Details from three different areas indicated in (a), showing the transmission of particular spot polymorphisms. O mark maternal fragments and ☐ mark paternal fragments. Arrows indicate non-transmitted polymorphic fragments.

The percentage of spot variants transmitted to the son was found to be 70% for both probe 33.6 and 33.15, rather than the expected 50%. This phenomenon could be due to clustering in the genome of VNTR loci or to the presence of one or more Hae III restriction sites in the minisatellites themselves (see above). In the latter case several variant spots could stem from the same minisatellite locus, thereby resulting in a number of spots being co-transmitted as minisatellite 'haplotypes' (Jeffreys *et al.*, 1986, see also 4.3.2 and 4.3.3).

The above shows the power of 2D DNA typing in following transmission of large numbers of spot variants from one generation to the next. To use this power effectively it is necessary as has already been mentioned above to increase the informativeness of the system by identifying spots as complementary alleles.

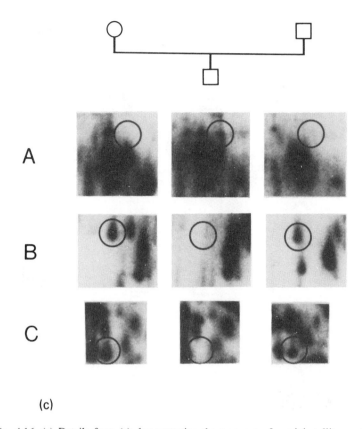

(c)

Fig. 4.16. (c) Details from (a) demonstrating the presence of a minisatellite containing
fragment in the son which is absent from the parents (circles in A) and two cases in which
both parents share a spot which is absent in the son (circles in B and C).

In Fig. 4.17 2D DNA typing patterns are shown of a human pedigree which is
part of the panel of human pedigrees of the CEPH consortium. These pedigrees
are the core of an international effort to map genetically the human genome (see
Chapter 2). Hence many different genetic marker loci have been typed in these
pedigrees by a consortium of cooperating molecular genetic laboratories (Dausset
et al., 1990, NIH/CEPH collaborative mapping group, 1992). By 2D DNA typing
of these pedigrees information can be obtained on the identity of spots by
observing co-segregation of particular spot variants with previously typed
locus-specific markers and with themselves. This is illustrated by Fig. 4.17
showing 2D DNA typing patterns of 2 parents from CEPH pedigree 1377 (Fig.
4.17a), a selected area from this pattern in 10 members from the pedigree (Fig.
4.17b) and a schematic representation of the segregation of spots from this region
(Fig. 4.17c). This approach, facilitated by the use of image processing techniques
(see also Chapter 3), will result in a 2D DNA typing database of fully
informative genetic markers as spots defined by (1) restriction enzyme, (2)
length, (3) melting temperature and (4) core probe. A multilaboratory

collaborative research project towards that aim is in progress. For the currently available 2D DNA typing system a larger sample size will compensate for the lower informativity. As such 2D DNA typing can be immediately applied in linkage studies without the need of first constructing a genetic linkage map. This is further illustrated below.

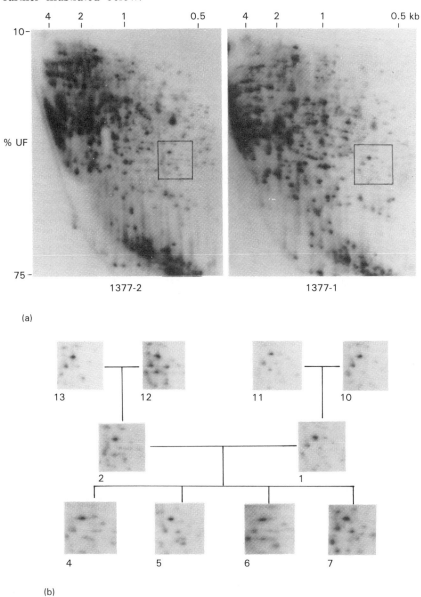

(a)

(b)

Fig. 4.17. (a) Two-dimensional DNA typing patterns obtained with minisatellite core probe 33.6 of two parents from CEPH pedigree 1377. (b) A selected area (see (a)) of the 2D DNA typing patterns of 10 members of CEPH pedigree 1377 (grandparents, parents and 4 children).

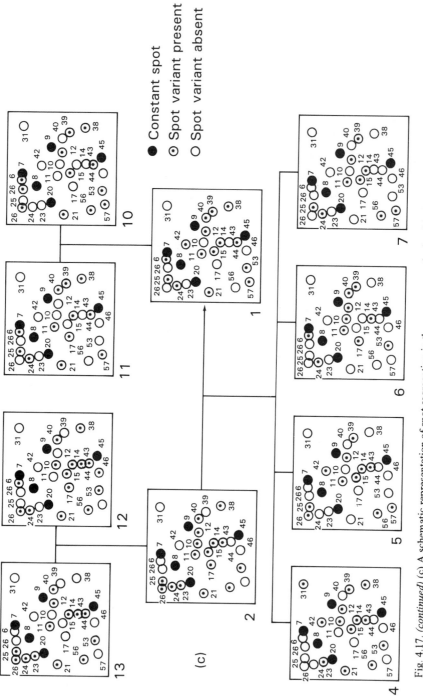

Fig. 4.17. *(continued)* (c) A schematic representation of spot segregation in the same area as depicted in (a). Comparison of the patterns and scoring of the spots was done by eye.

4.3.4.2 Animals

Two-dimensional DNA typing patterns comparable with those for humans have been obtained for mice and rats (see, for example, Chapter 3). As for humans, for the mouse a high-density genetic map has been obtained (Dietrich *et al.*, 1992). For most species this is not yet the case. A unique advantage of the 2D DNA typing system is the large number of alleles that can be analysed simultaneously by a given core probe. This feature no longer necessitates the construction of a genetic map consisting of large numbers of locus-specific probes/primer sets. Obviously, this is not an advantage when a high-density map is already available, such as for the human and mouse genome. However, for other species the construction of a genome map of some density is an investment that is now no longer necessary. The development of such informative markers involves several man-years of screening genomic libraries at random. Moreover, with the random isolation approach it becomes increasingly difficult to isolate more markers, especially from regions with a low marker density. With 2D DNA typing, only when one or more interesting spots have been observed there is the need to develop these into locus-specific probes. This can be accomplished by following a recently developed PCR-based protocol for the direct isolation of the micro- or minisatellite flanking sequences from the 2D gel (de Leeuw *et al.*, in preparation; see also above). The 2D DNA typing method is therefore immediately available for any species with a genome harbouring polymorphic micro- and minisatellite loci (see Table 4.3 for a list of suitable probes). Gene-mapping projects for such species can start immediately without any prior investments in the form of genomic map construction. Indeed, while conducting a gene search with 2D DNA typing such a map is automatically constructed by studying co-segregation of spots in the pedigrees of interest (see above).

Table 4.3. Usefulness of selected micro- and minisatellite core probes for genetic studies in various species[a]

	Species										
Probe	Human[b]	Ape	Cow	Horse	Mouse	Rat	Pig	Pigeon	Parrot	Tomato	Yeast
33.15	A	A	C	A	A	A	A	A	A	A	A
HBV-3	A	A	B	B	F	A	A	A	B	F	F
YNZ-22	A	A	B	F	C	A	B	A	A	B	F
M13	A	A	A	A	F	F	F	A	A	F	F
INS	A	A	A	F	E	A	A	A	A	A	E
HYCO	A	A	C	E	A	B	A	A	A	C	F
33.6	A	A	A	A	A	A	A	A	A	A	F
RECO	A	F	D	F	A	A	F	F	F	F	F
(GACA)$_n$	B	A	B	B	F	A	A	B	A	F	F
(CAC)$_n$	A	A	A	A	A	A	A	A	A	F	A
(AGC)$_n$	A	A	C	B	D	A	F	B	B	B	A
(TCC)$_n$	A	F	A	F	F	F	F	F	F	F	F

[a]Based on Southern blot hybridization analysis (see also Fig. 2.7).
[b]A, many polymorphic bands (≥10 bands); B, few polymorphic bands (<10 bands); C, non-interpretable owing to large number of bands; D, no banding pattern; E, non-polymorphic banding pattern; F, not tested.

The strategy described above for the analysis of human DNA also pertains to animal DNA. This is illustrated by Fig. 4.18 showing core probe hybridization patterns obtained with cattle genomic DNA (*Bos taurus*). Spot patterns, comparable with those found for human DNA, were observed, although with a lower number of spots, i.e. 430 and 420 with 33.6 and $(CAC)_n$, respectively. In the 2D DNA typing pattern of cattle DNA obtained with the core probe 33.15 an intense background hybridization can be observed below the diagonal of fragments that are separated by size but not yet melted. Currently, we do not know the cause of this hybridization phenomenon but it might be related to the cross-hybridization of this core probe to the bovine satellite I repeat sequence (Jeffreys, 1985). Interestingly, however, above the diagonal of size-separated but not melted fragments a reasonably clear pattern of about 200 spots can be observed, which might represent true minisatellites with homology to the 33.15 sequence.

Table 4.4 summarizes the number of spots and the percentages of variant spots for several core probes obtained with cattle genomic DNA. Overall, there do not seem to be considerable differences with the human situation (Table 4.2). For cattle, the overlap in spot positions has not yet been accurately determined, but preliminary data suggest rather identical values and mainly restricted to particular probes, such as HBV-1 and $(TCC)_n$. This overlap was also observed when evaluating these core probes for one-dimensional DNA profiling analysis of cattle (Trommelen *et al.*, 1993)

Table 4.4. Core probe characteristics in 2-D DNA typing of cattle[a]

Probe	Number of spots	Spot variants (%)
33.6	172	27
33.15	247	27
INS[b]	245	25
HBV-1[c]	109	17
ZETA	195	13
HBV-3	75	ND[d]
$(GATA)_n$	50	ND[d]
$(CAC)_n$	326	18
$(TCC)_n$[e]	156	13
TELO	39	ND[d]
Total	1614	

[a]Based on the analysis of at least two unrelated individuals.
[b]INS has 18% of its spots in common with CAC.
[c]HBV-1 has 60% of its spots in common with 33.6.
[d]ND = not determined.
[e]$(TCC)_n$ has 32% of its spots in common with 33.6.

The usefulness of 2D DNA typing in finding markers for genetic traits by taking the linkage approach is illustrated by our studies on the red/black phenotype in cattle (te Meerman *et al.*, 1993). A pedigree with 1 sire and 12 dams was used, having 16 offspring, shown in Fig. 4.19. The red/black

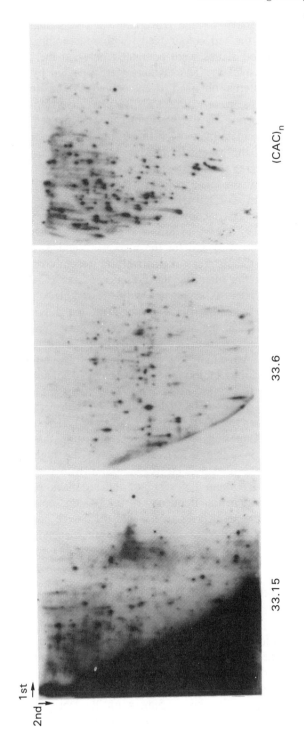

Fig. 4.18. Two-dimensional DNA typing patterns of cattle obtained from Hae III digested genomic DNA with core probes 33.15, 33.6 and $(CAC)_n$ (from Uitterlinden et al., 1991b).

phenotype is recessive for the red colour, and was analysed by crossing a sire heterozygous for the red colour factor with red (homozygous) dams. Two-point and three-point lod scores were computed between the red/black phenotype and five selected 2D markers. The markers were selected by eye examination of 2D DNA typing patterns obtained from pooled DNAs from the red and black offspring. (Patterns were confirmed by also analysing the individual DNAs.) Those spots occurring in the black but not in the red pools (Fig. 4.20) were considered as candidate markers. It turned out that the fourth probe used, $(CAC)_n$, provided most of the candidate markers, showing co-segregation with the colour factor. The pedigree is informative for linkage because the design makes it possible to follow the male meiosis for 16 offspring. The data were analysed with the program GRONLOD (te Meerman, 1991), a program for linkage analysis where single unknown alleles are permitted. Segregation in the cattle pedigree (Fig. 4.19) is easy to follow. The sire is evidently heterozygous for red (number 1 in Table 4.5) and for the markers 2 and 3. Dam truus 6 is probably homozygous for the marker (alternatively, a recombinant in the sire may have occurred), and in the offspring the presence of the $(CAC)_n$ allele visible in the 2D DNA typing analysis segregates with the red/black phenotype, such that the allele is in phase with the 'black' factor. This explains the excess of the alleles in the black cows and the absence in the red ones. No recombinant is required to explain the segregation of the marker. Using one meiosis for phase determination, a maximum lod score for a fully informative marker of 15 log(2)=4.5 is possible in this pedigree design. The auto-lod score of the coat colour locus has such a lod score. The 2D markers are less informative because only the presence of a single allele is detected with certainty. This appears from the auto-lod scores of 0.99 to 3.47 (Table 4.5). The matrix of two-point lod scores, including auto-lod scores and recombination rates for the maximum lod score, is shown in Table 4.5. It appears that the highest lod score of 2.6 is present between the marker 3 (segregation shown in Fig. 4.19) and the red/black phenotype. The maximum lod score of the red/black phenotype with marker 2 is 0.95 at a recombination frequency of 0.15. Three-point generalized lod scores with recombination frequencies of 0.0 and 0.18 were computed for the red/black phenotype and the markers 2 and 3. The results depended on the assumed locus ordering:

	lod score
order 1,3,2 and 3,1,2	3.5
order 3,2,1	3.0

The order information is of course not yet very reliable. All recombinants are observed in the father, which is not surprising because the mothers are all homozygous red and only those with two offspring could theoretically show recombinants between marker alleles.

These results illustrate the usefulness of 2D DNA typing in rapidly finding linkage between an economically important monogenic trait and a VNTR allele.

The spot co-segregating with the colour factor is at present being isolated from the gel to be used as a locus-specific marker.

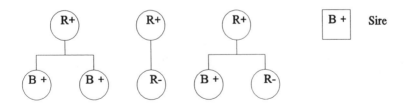

+ indicates presence marker allele
- indicates absence marker allele
B indicates black phenotype
R indicates red phenotype

truus 6

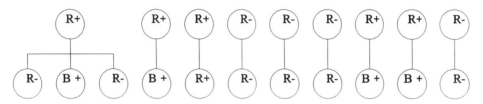

Fig. 4.19. The cattle pedigree consisting of 1 sire, 12 dams and 16 offspring, which was used to identify co-segregation of spots with the red/black phenotype. The segregation of the best linked marker (marker 3 detected by $(CAC)_n$; see Table 4.5) is indicated ('+' indicates the presence and '−' the absence of the spot).

Table 4.5. Lod score table for the coat-colour locus in cattle with $(CAC)_n$ 2D spots[a]

	red/black	2	3	4	5	6
red/black	**4.5**	0.15	0.0	0.3	0.45	0.2
2	0.95	**3.35**	0.0	0.5	0.0	0.2
3	2.57	1.48	**2.70**	0.5	0.5	0.2
4	0.14	0.0	0.0	**3.47**	0.2	0.15
5	0.0	0.14	0.0	0.10	**0.99**	0.5
6	0.59	0.30	0.20	0.52	0.0	**2.75**

[a]Auto-lodscores are on the diagonal (printed in bold) and maximum two-point lod scores between the red/black phenotype and five 2D markers (2–6) are indicated below the diagonal. Above the diagonal the recombination rates at maximum lod scores are presented.

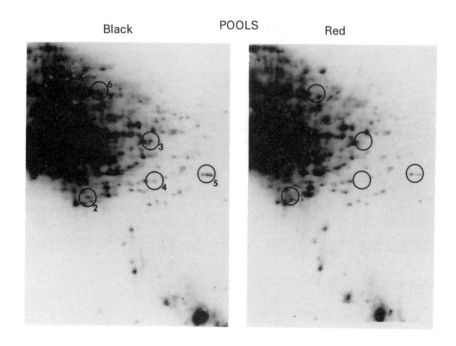

Black   POOLS   Red

(a)

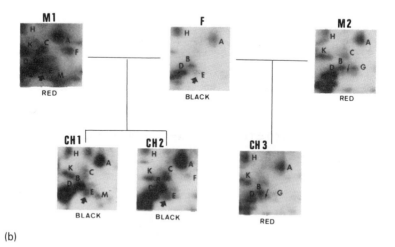

(b)

Fig. 4.20. (a) Two-dimensional DNA typing patterns obtained with microsatellite core probe
$(CAC)_n$ of the black and red pool. Candidate 2D DNA typing markers (i.e. present in the
black, but not in the red pool) are indicated. It turned out that spot 3 had the highest lod score.
(b) Details from (a), showing the segregation of spot 3 in the offspring of the site (F) and two
dams (M1 and M2). A thick arrow indicates presence of the $(CAC)_n$ spot, a thin arrow indicates
absence of the spot. A–M indicate other spots for reference (from te Meerman *et al.*, 1993).

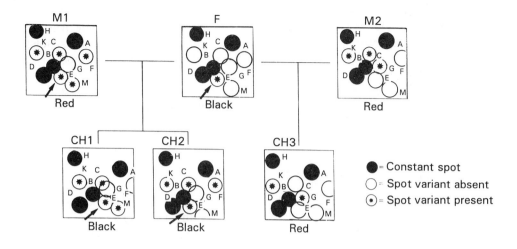

(c)

Fig. 4.20 *(continued)*. (c) Schematic representation of (b). The arrow indicates spot 3.

4.3.4.3 Plants

In principle 2D DNA typing can be applied for genetic studies on plants. However, for plant species with very large genomes (for example, lilies) hybridization analysis becomes very insensitive. Also, when working with inbred lines suitable (polymorphic) marker systems can be selected, which is in contrast to the situation with humans; suitable crosses can not be selected at will. Therefore, in plants diallelic RFLP markers and, more recently, RAPD markers have been the method of choice. In plant genetic studies the accent is very heavily placed on cost efficiency of the analysis.

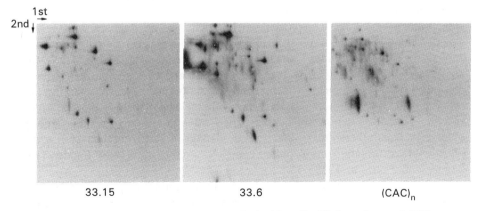

Fig. 4.21. 2D DNA typing patterns of tomato obtained from Hae III digested genomic DNA after hybridization with core probes 33.15, 33.6 and $(CAC)_n$ (from Uitterlinden *et al.*, 1991b).

Nevertheless, 2D DNA typing could play a role also in plant genetic studies and some examples are shown here. Fig. 4.21 shows 2D DNA typing patterns of tomato (*Lycopersicum esculentum*) consisting of 89, 117, and 140 spots (total 346) for 33.15, 33.6 and $(CAC)_n$, respectively. These patterns are much less complex than the patterns obtained for humans or for cattle. Furthermore, considerable overlap (50%) in the spots detected was observed among these three core probes in tomato, indicating extensive homology of the sequences detected with all three probes.

4.4 ASSOCIATION STUDIES

Two-dimensional DNA typing can be used for mapping simple Mendelian traits in segregating populations as was illustrated by the red factor in cattle (see 4.3.4.2). However, the method may be even more useful in identifying the genomic regions that contain genes involved in quantitative traits. Examples of such traits are economically important characteristics such as milk production in cattle or certain complex diseases in humans. The power for detecting such genomic regions is, however, much lower than for the relatively simple genetic traits. The usefulness of any method, including 2D DNA typing, for addressing such complex traits critically depends on the power that can be obtained.

Monofactorial diseases show co-segregation with marker alleles in pedigrees. Power analysis for discovering marker loci linked to disease genes is in principle straightforward. Depending on the pedigree structure, more or less linkage information is obtained, and a power curve can be computed depending on the recombination fraction between the marker and the disease and the degree of heterozygosity displayed by the marker. If a sufficiently large sample is analysed, every monofactorial disease will eventually be mapped, and finally the gene will be found. This paradigm has been very successful over recent years, although for humans lack of suitable pedigrees for analysis make precise location through a linkage strategy often difficult.

In principle the linkage approach can be applied to all diseases with a simple genetic component or more generally to any monogenetic trait in humans, animals or plants, provided suitable pedigree material is available. However, a number of problems may arise (e.g. reduced penetrance, aetiologic heterogeneity, presence of phenocopies) which can be surmounted when working with experimental animals or plants, but not with humans which cannot be crossed at will. The most serious difficulties arise when a disease or other trait has several loci and therefore several linked markers at different loci (for reviews of possible difficulties in mapping complex genetic traits see: Lander, 1988; Lander and Botstein, 1989).

In multifactorial diseases alleles at a number of different loci interact in an additive fashion to determine (in combination with many different environmental exposures) an individual's risk of a trait or disease. Unfortunately, this is also the most frequent form in which genetic traits become manifest in the human population as well as in economically important animals or plants. In humans, multifactorial diseases (e.g. heart diseases, cancers, diabetes, arthritis, high blood pressure) are a set of diseases for which the genetic model is unknown. This implies that power studies are possible only while assuming certain models of inheritance (Clerget-Darpoux and Bonaïti-Pellié, 1992).

All genetic models for complex diseases will specify that the presence of certain variants of genes increases or decreases the probability of a disease. Interaction with the environment is an additional factor for a complete understanding of such diseases. Parameters are the number of genes at different loci and the number of alleles of genes involved. Genetic variation observed at the DNA level can be found correlated with phenotypic variation as a result of several mechanisms.

(1) If allelic variation within a gene is found, it will lead to association between the presence or absence of an allele and a phenotype.

(2) A linkage disequilibrium between an anonymous marker allele and a disease allele will lead to an increased or decreased allele frequency associated with a phenotype.

(3) Linkage between anonymous markers and disease genes.

(4) Random results, arising from multiple testing and chance.

The interpretation of associations is quite difficult as appears from the many known associations between disease and HLA alleles (Segall and Bach, 1990). There is not yet a complete explanation for such associations, probably because the HLA alleles are co-factors in a multifactorial system. It would however be quite important to know more of such co-factors, because only when several associations are known can a theory be made about the genetic contribution to chronic disease.

Two-dimensional DNA typing has the potential to reveal thousands of polymorphisms at the DNA level, by direct detection of the absence or presence of alleles. Given the huge reservoir of different micro- and minisatellite core probes (Uitterlinden *et al.*, 1991a), a marker density of one million basepairs or less can theoretically be achieved by re-hybridizations. Therefore, the application of 2D DNA typing for association studies should be oriented at a complete genome search, with the potential to reveal genes directly (when the marker is in the gene itself) and through linkage disequilibrium. This may result in one or more spot variants of interest being isolated and developed into a fully informative locus-specific marker. With the aid of this fully informative marker more information can be obtained from the same data set, to serve as a confirmation of initial results.

The statistical power of such a study will never reach 100%, even if the population material is huge. The reason is that the probability of direct detection of a susceptibility gene is very low (except when screening cDNAs; see Chapter 5) and the probability that a suitable linkage disequilibrium is present is also below 100%. Markers should also have enough polymorphism in order to obtain sufficient data. A linkage disequilibrium does in fact to a great extent depend on the occurrence of a rare mutation of the gene in coupling with a rare allele or haplotype, as has been the case in cystic fibrosis with the XV2c/KM19 haplotype disequilibrium (Maciejko *et al.*, 1989).

If an ordinary significance level of 5% is used, and allowance is made for 1000-fold multiple testing, the significance threshold of any test should, using Bonferroni's correction for multiple testing, be 5.0×10^{-5}. Fig. 4.22 shows a graph where the required sample size for 95% power to detect an increase of the allele frequency can be found as function of the allele frequency and the factor with which the allele frequency is increased. The factor increase varies between 1.25 (25% increase) and 4 (400% increase). It is clear that, the larger the increase, the lower is the required sample size to obtain the indicated power. The lines indicate the natural logarithm of the sample size.

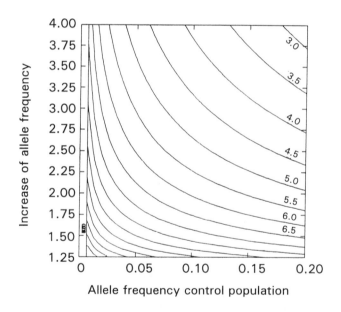

Fig. 4.22. Required sample size to detect increases in allele frequency in patients as a function of its normal frequency in the population. Sample size (x) can be calculated by $x = e^y$, where y is indicated in the figure above the lines.

In summary, 2D DNA typing could become the method of choice in future attempts to localize genes involved in complex traits through association studies. However, it should be realized that associations due to identity by descent are not detectable in case of a large number of independently originating disease mutations. Nevertheless, 2D DNA typing would for the first time open up the possibility of conducting association studies systematically, without the need to rely on candidate genes or, even worse, to embark on a fishing expedition by testing as many markers as possible at random.

4.5 GENETIC ANALYSIS OF CANCER

It has been shown that a number of different chromosomal regions display allelic imbalance (gain or loss of one of the alleles at a locus) in different types of tumours (for reviews, see Ponder, 1988; Fearon and Vogelstein, 1990). Especially the loss of chromosome segments is a frequent occurrence in tumours. The search for such losses has become relatively easy with the availability of more and more polymorphic markers (Vogelstein et al., 1985). This allows the loss of a chromosomal fragment to be detected as a so-called loss of heterozygosity (LOH) on particular loci (Ponder, 1988).

An example is breast cancer in which a number of different chromosome arms were shown to have allelic imbalance (Mackay et al., 1988; Ali et al., 1989; Devilee et al., 1989). The most notable imbalance has been shown to be the short arm of chromosome 17, with imbalance in 50–60% of the informative cases (Devilee et al., 1990). One reason for a non-random distribution of allelic imbalance may be the presence of susceptibility genes (e.g. oncogenes, tumour suppressor genes) at the loci of the chromosomal regions involved. At least one tumour suppressor gene often involved in breast cancer, the TP53 gene (for review, see Levine, 1992), is located at 17p (Isobe et al., 1986; Nigro et al., 1989; Baker et al., 1990). Studying early-onset breast cancer kindreds, taking the linkage approach, Hall et al. (1990) obtained evidence for the presence of a breast cancer susceptibility gene (BRCA1) in a 50 million basepair region of chromosome 17 (17q21). Subsequent analysis with more markers in more families has narrowed down the location of this autosomal dominant susceptibility gene to a few million basepairs (Chamberlain et al., 1993).

Furthermore, chromosomal regions on 1q and 6q have also been frequently involved in both cytogenetic changes and in studies of allelic imbalance (Mitelman, 1991; Sager, 1989). For example, loss of heterozygosity on chromosome 1p34-p36 was most frequent in patients with strong family history of breast cancer (McGuire and Naylor, 1989). In addition, chromosome arm 13q, the location of the tumour suppressor gene RB1, displays allelic imbalance in 20–30% of the cases (T'Ang et al., 1988). Also in other major forms of cancer consistent DNA sequence deletions have been observed. For example, in carcinoma of the lung, such deletions have been found, most notably at the chromosomal region 3p21 (Kok et al., 1987; Weston et al., 1989). Multiple genetic changes have also been described for colorectal cancer, including deletions involving chromosomes 5, 17 and 18 (Delattre et al., 1989). Also here p53 mutations were suggested by the data to be involved in colorectal neoplasia (Baker et al., 1989), while attempts are being made to define a subset of genetic alterations that can be used clinically to help assess prognosis (Kern et al., 1989). Recently, close linkage of the gene for hereditary non-polyposis colorectal cancer (HNPCC), a major form of colorectal cancer predisposition, to anonymous microsatellite markers on chromosome 2 was demonstrated in HNPCC kindreds (Peltomäki *et al.*, 1993; Aaltonen *et al.*, 1993; Ionov *et al.*, 1993). Interestingly, rather than showing loss of heterozygosity for chromosome 2 markers, most familiar and some of the sporadic cancers showed changes in microsatellite alleles from various chromosomes, including chromosome 2 (Aaltonen *et al.*, 1993; Thibodeau *et al.*, 1993). This suggests that genetic instability, in particular that seen at micro- and minisatellite loci, is a component of the familiar cancer phenotype (Aaltonen *et al.*, 1993). It should be noted that instability at microsatellites can be involved in human disease. Expansion of trinucleotide repeats has been identified as the basis of disease in several inherited disorders (see Chapter 2) and similar events may be important in cancer.

Thus, the studies of allelic imbalance are based on the use of DNA markers which allows one to identify the lost allele in the tumour tissue. The allele then represents a larger area of the chromosome in which genes crucial to the arisal and development of the tumour are likely to be located. This opens up possibilities of designing DNA-based tests to predict tumour behaviour in individual patients. Information on all possible genetic changes in a given tumour and the finding of significant associations between such changes and clinicopathological data could provide a set of prognostic indicators.

Genome searches to identify chromosomal regions altered in the tumour tissue have consisted of the serial application of different locus-specific probes, the number and location of which determine the efficiency and accuracy of the search. To increase the number of informative cases, locus-specific VNTR probes have been used in several studies. One example is pYNZ22 (Nakamura *et al.*, 1988b) which detects a VNTR locus on chromosome 17p, close to which a region is present that has been assumed to play a role in the development of breast cancer (Sato *et al.*, 1990). Another example involves the RB1 gene in which VNTRs are residing within introns 17 (Wiggs *et al.*, 1988) and 21 (Yandell and Dryja, 1989).

The application of micro- and minisatellite core probes greatly increases the number of sites that can be detected simultaneously in a search for alterations in the tumour genome. Examples of 1D DNA fingerprint analysis of tumours include the detection of deletions and amplifications in many different types of tumours (Thein *et al.*, 1987; Armour *et al.*, 1989). By means of 2D DNA typing the number of sites in the genome which can be simultaneously analysed for variations can be increased to several thousands, simply by hybridization analysis with many different core probes.

The 2D DNA typing method can contribute to the analysis of the tumour genome in two ways. First, the total extent of genomic alterations can be determined, which by itself could have prognostic relevance (Kern *et al.*, 1989; Sato *et al.*, 1991). Second, the DNA fragments corresponding to particular changes in the tumour genome can be isolated from the 2D DNA typing gel and developed into locus-specific markers. These can then be subsequently used to identify the chromosomal regions involved in the genomic alterations. This could lead to the identification of candidate tumour-related genes.

These two applications are illustrated by 2D DNA typing of a small series of breast cancers for genomic instabilities (Verwest *et al.*, 1994). Comparison of tumour and normal DNA of 11 breast cancer patients, using 2D DNA typing with four core probes, revealed a considerable number of genomic alterations. On average 863 spots were scored for all probes together per individual after correction for spot overlap. Fig. 4.23 shows two-dimensional spot patterns of tumour and normal (white blood cell) DNA from patient 43 after hybridization with core probes 33.15 and TELO. In all patients differences were observed (on average 2.1%; range 1.1–5.2%) between tumour and normal tissue. Some details

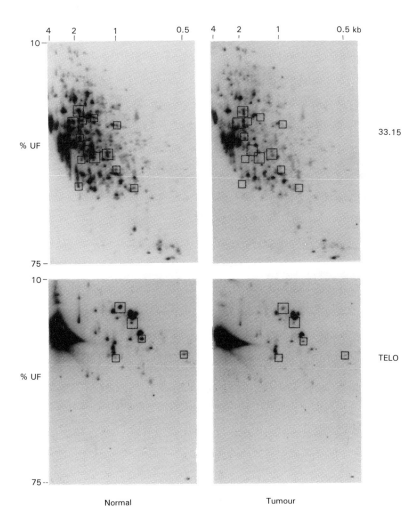

Fig. 4.23. Two-dimensional DNA typing patterns of breast tumour DNA (tumour) and WBC DNA (normal) from patient 43 after hybridization with core probes 33.15 and TELO. Squares indicate spot differences between normal and tumour DNA (from Verwest *et al.*, 1994).

are shown in Fig. 4.24. Most of the changes observed were deletions (absence of spots in the tumour) or amplifications (spots of higher intensity in the tumour). About 10% of the genomic changes detected appeared to occur in the tumours of more than one patient. The variation among patients is illustrated by Fig. 4.25, in which the genetic changes detected by Southern analysis were combined with those obtained by 2D DNA typing. The figure indicates that the micro- and minisatellite core probes used have the same tendency but detect different spectra among patients with respect to the frequency of changes detected in a given patient. In view of the small number of patients no clinically relevant correlations could as yet be made. However, it is clear that 2D DNA typing is potentially very powerful in studying the prognostic value of genetic instabilities in cancer.

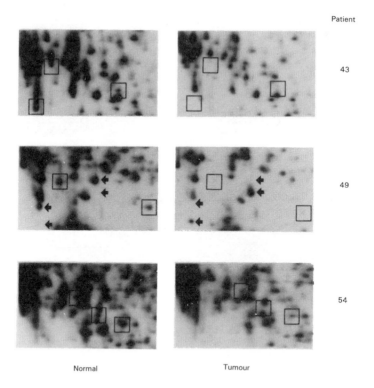

Fig. 4.24. Details taken from 2D DNA typing patterns of normal vs tumour DNAs from three different patients after hybridization with probe 33.6. Squares indicate decreases in intensity of a spot and arrows indicate shifts in position of a spot between tumour and normal DNAs (from Verwest *et al.*, 1994).

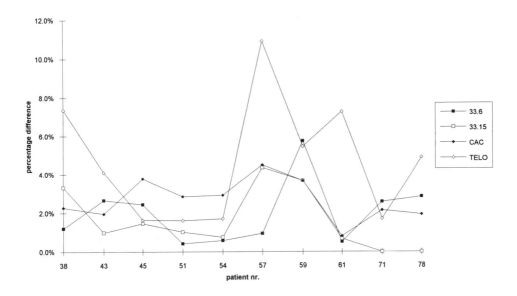

Fig. 4.25. Percentage of genetic changes of the total number of fragments detected by both 2D DNA typing and Southern blot hybridization analysis, in breast tumours of different patients using each of four different core probes, i.e. 33.6, 33.15, $(CAC)_n$, and TELO (from Verwest et al., 1994).

Fig. 4.26 shows the gel positions of all differences found in more than one patient. As yet the number of patients tested is too small to draw any conclusions. However, the observation of specific changes in groups of patients suggests that the genetic regions involved could be of prognostic value. It should be noted that spot variants found to be on the same isotherm in the denaturing gradient (second dimension) could be alleles from the same locus (see 4.3.2). These (polymorphic) spot variants (Fig. 4.26) are prime candidates for follow-up studies. Indeed, these spots have been isolated directly from the gel and are at present being tested as locus-specific probes on larger numbers of primary breast tumours for their suitability as prognostic indicators. The polymorphic nature of these probes allows their use in studying genetic susceptibility of breast cancer. By now there is ample evidence that particular regions of genomic instability in breast cancer may point the way to genes that may be causally related to the development of breast cancer. The availability of ample genetic markers will greatly facilitate the identification and isolation of genes determining tumour behaviour and susceptibility.

On the basis of the initial data on breast tumours presented above (Verwest et al., 1993) a logical follow-up would be to analyse comprehensively several of the most frequent human cancers (e.g. breast, lung, colon cancer) for instabilities in their genome. A database of these mutations could then be compiled and their corresponding genomic regions systematically searched for candidate genes.

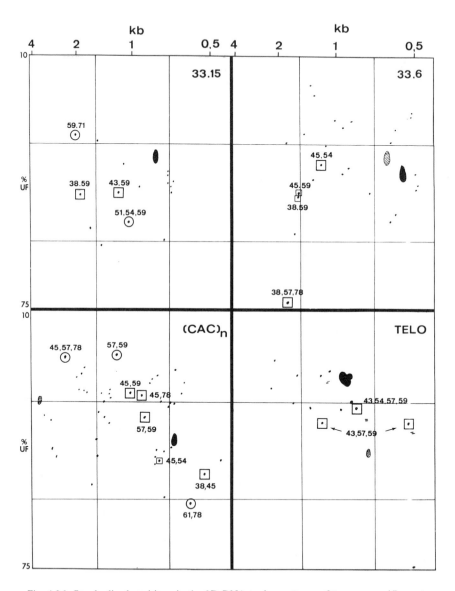

Fig. 4.26. Standardized positions in the 2D DNA typing patterns of tumour-specific spot variants detected in at least two different patients using four different core probes (33.15, 33.6, (CAC)$_n$ and TELO). For each core probe used the standardized grid and the constant spots used to compare different gels are shown. Circles indicate gains/amplifications. Squares indicate spot losses. Numbers adjacent to the circles/squares indicate the patient number (from Verwest *et al.*, 1994).

4.6 MUTATION ANALYSIS

Current approaches for biomonitoring humans thought to be at risk for DNA damage include mutation analysis at selectable genes such as HPRT (Branda *et al.*, 1993) and MHC (Grist *et al.*, 1992). Although the results thus far obtained with such systems have greatly contributed to our understanding of the mechanisms underlying mutagenesis *in vivo*, these methods nevertheless suffer from the disadvantages that (a) selection of cells in culture is involved, (b) because of their inherent stability coding genes could be insensitive indicators for permanent genetic damage, and (c) mutations are measured at only a single site in the genome (see, for example, Neel *et al.*, 1993). An alternative would be to look directly (i.e. without selection) at large numbers of naturally unstable (hypermutable) mutational target sequences spread over the genome. A category of naturally present DNA sequences that offers such a possibility is micro- and minisatellites.

The human genome is estimated to contain several thousands of micro- and minisatellite DNA sequences more or less evenly spread over the different chromosomes (Chapter 2). Together these sequences roughly constitute about 1% of the total number of basepairs. It has been demonstrated that a fraction of them (the so-called variable number of tandem repeats or VNTRs) represent a relatively 'unstable' part of the human genome displaying both somatic instability, such as in tumours (Thein *et al.*, 1987; Armour *et al.*, 1989; see above), and germ line instability (Jeffreys *et al.*, 1988; Nürnberg *et al.*, 1989). The latter is considered to be responsible for the extensive allelic variation in the human population of these sequences.

By using micro- and minisatellite core probes, in Southern blot hybridization analysis, so-called genomic DNA fingerprints can be generated (Jeffreys *et al.*, 1985a and 1991a; see Chapter 2). Each core probe generally detects DNA sequences at hundreds of different loci, of which only few (the largest) can be resolved by one-dimensional DNA fingerprint analysis. By employing 2D DNA typing, which is capable of resolving virtually all members belonging to a particular set of micro- and minisatellites (with exception of the largest), the genome can be 'scanned' for variations in micro- and minisatellite DNA sequences.

The approach for the measurement of human DNA mutation rates by 2D DNA typing could offer a complementary measuring device in addition to current mutation detection systems. An important issue in this respect is whether or not micro- and minisatellite regions are subject to similar types of mutation processes as other (e.g. coding) regions in the genome. One way of evaluating this aspect is to analyse induced mutations in cloned cells caused by agents which are known to cause mutations in for example the HPRT gene.

DNA fingerprinting has been applied in mutation studies (Honma *et al.*, 1993; Kikuno *et al.*, 1992; Dubrova *et al.*, 1993). The strategy employed usually involves the Southern analysis of a number of subclones from treated as well as from untreated (control) cell lines, using a panel of minisatellite probes. As expected, relatively high spontaneous mutation frequencies were found, in the order of 10^{-4}–10^{-3} (Honma *et al.*, 1993). Also, it was found that treatment with known mutagens increased the mutation frequency (Kikuno *et al.*, 1992). More recently, *in vivo* treatment of mice with ionizing radiation showed an 1.7 times increase in the mutation frequency at minisatellite loci from 0.0085 to 0.0147 per band per offspring (total offspring analyzed was 232), as measured by one-dimensional DNA fingerprinting analysis of offspring of treated animals (Dubrova *et al.*, 1993).

In some preliminary studies 2D DNA typing has been used to study mutation frequencies in fibroblasts from the skin of old vs young rats (Slagboom *et al.*, 1991). The results of this study confirm the results from the studies mentioned above; micro- and minisatellite loci have an average mutation rate between 10^{-4} and 10^{-3}. Thus far, large-scale 2D DNA typing studies on treated cells have not yet been performed.

In interpreting the results obtained on mutation frequencies at micro- and minisatellite loci it should be realized that the sequences detected by core probes comprise an array of different loci. This might be the result of differences in sequence composition of the repeat unit of which a locus is composed, the length of alleles of the locus, the location in coding vs noncoding regions in the genome, and the location on the chromosomes (e.g. subtelomeric vs interstitial). Some loci (the non-VNTR loci) are known to be almost completely 'silent' and have a mutation frequency very close or identical to that of other DNA sequences (Jeffreys *et al.*, 1988; Nürnberg *et al.*, 1989). Some other loci (the VNTR loci) are known to be mutable within a very broad range of mutation frequencies but sometimes to an extreme extent. For example the human VNTR locus detected by probe pMS 1 (D1S8) is known from pedigree studies to have a (meiotic) mutation frequency of 1/20, meaning that 5% of the gametes of an individual have a mutated allele at this locus (Wong *et al.*, 1987; Jeffreys *et al.*, 1988). From previous studies (Nakamura *et al.*, 1987a; Uitterlinden *et al.*, 1989a; Uitterlinden *et al.*, 1991a; Armour *et al.*, 1989) it has been estimated that approximately 25–40% of all micro- and minisatellite containing restriction fragments detected by one core probe are derived from VNTR loci. It should be noted that among the VNTR loci analysed so far for population polymorphism and mutation events, the ones with long (>2 kb) alleles are the most polymorphic and mutable (Jeffreys *et al.*, 1985a, 1988).

4.7 CONCLUDING REMARKS

Two-dimensional separations of restriction-enzyme-digested genomic DNAs from various organisms, followed either by direct detection with sensitive stains or by hybridization analysis with repeat element probes, appears to be a powerful genome scanning system. The examples given above illustrate a range of applications, varying from molecular taxonomy for microorganisms to systematically searching the tumour genome for clinically relevant alterations. Perhaps the major application of the system will be found in the identification of unknown genes, either through the linkage approach when relatively simple genetic traits are involved or by the detection of marker associations with genetically complex traits. Although in the above attention has been focused on humans, it is not unlikely that initially 2D DNA typing will be used most frequently in breeding studies on economically important animals and plants. This has to do with the relative lack of highly polymorphic markers for genomes other than human and mouse. Indeed, even for animals of relatively high value, such as cattle, high-resolution genetic maps are not available (Womack, 1992). Therefore, the possibility of immediately starting to analyse marker–QTL relationships in crosses between breeds at no initial costs is an attractive one. We anticipate that in the near future 2D DNA typing databases will become available with locus information on most spots detected by a set of core probes for a given species. In combination with a stack of hybridization patterns for each individual in a given pedigree, such a database will allow the automatic processing of the data which can then immediately be entered in suitable linkage programs. By coupling to genomic databases with information on physical map, sequence map etc., any spot that proved to be linked to a particular phenotype could then be conveniently retrieved from a YAC or cosmid clone in a library. To this end the development of image-processing and genetic analysis software is at present in progress and it can be foreseen that automated instrumentation will follow suit (see Chapter 3 and Mullaart *et al.*, 1993)).

5

Future developments

5.1 INTRODUCTION

The application of 2D DNA typing to analyse a selected set of DNA samples will result in the identification of particular spots of interest. As was illustrated in the previous chapter these can include spots showing close genetic linkage or association with a genetic trait as identified after pedigree or population analysis, or spots being lost, gained or amplified in tumour DNA in comparison with normal DNA or in normal DNA after exposure to mutagenic agents. The identification of the particular spot variant as an allele of a particular locus on one of the chromosomes is key in the subsequent isolation of the gene in question by so-called positional cloning. Such a physical mapping of the spot variant can be achieved by two methods.

First, the spot variant can itself be isolated from the 2D gel and developed into a locus-specific probe (de Leeuw *et al.*, in preparation; see also Chapter 4). Second, the chromosomal location of a particular spot variant can be determined by following its transmission in pedigrees in which the transmission of many other markers with known chromosomal location is documented. By observing co-segregation of the spot variant with one or more of the markers already typed in the reference set of pedigrees (for example, the CEPH panel; see Chapters 2 and 4), the location of the corresponding micro- or minisatellite locus can be inferred from the observed genetic linkage. This procedure has been described in Chapter 4 and is at present being followed to generate the first 2D DNA typing human genetic linkage map (Mullaart *et al.*, in preparation; see also te Meerman *et al.*, 1993).

Once a locus-specific probe is available for a particular spot variant identified by 2D DNA typing, the chromosomal region in which it is located can be searched for additional markers to refine genetic mapping and for candidate genes likely to be involved in the biological process which was studied by 2D DNA typing of the samples. Also in these important subsequent steps in the positional cloning process 2D DNA typing, albeit in very different formats, can be applied to achieve efficient scanning of the genome at an increasing resolution down to the detection of a basepair mutation. This stratified manner of genome analysis is depicted in Fig. 5.1. First, the genetic mapping can be refined and even combined with physical mapping by the exploitation of other sets of very frequently occurring repetitive elements, i.e. interrepeat PCR analysis. Second, the identification of the gene encoding the trait of interest among the potentially many candidates within the marker-defined chromosomal region can be facilitated by 2D DNA typing of the corresponding (m)RNA transcripts (2D cDNA typing). Third, the final identification of the disease-causing mutation(s) among the many possible sites in the gene is simplified by simultaneous scanning of all exons with high accuracy for sequence variation (2D gene typing).

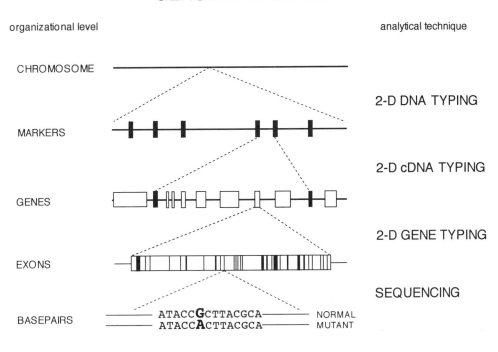

Fig. 5.1. Schematic depiction of a stratified manner for genome analysis and application of possible analytical techniques.

Of these three 2D DNA typing modalities none has yet reached the stage of experimental maturity that is necessary for application. The following is therefore merely a discussion of the concepts underlying these three modalities which in themselves are a logical continuation of the parallel approach to genome analysis.

5.2 POSITIONAL 2D DNA TYPING: THE CONCERTED USE OF GENETIC AND PHYSICAL MAPPING

For genetic fine mapping by 2D DNA typing, sequences bordered by repetitive elements, such as Alu I or Kpn I, are prime candidates. By the so-called interrepeat (ir) PCR approach human genomic sequences can be obtained not only from total genomic DNA, but also from complex sources such as human–hamster somatic cell hybrid cell lines containing individual human chromosomes or parts thereof. This is illustrated in Fig. 5.2 for inter-Alu sequences obtained from a panel of somatic cell hybrids containing (parts of) the human chromosome 21. By comparing the electrophoretic (size) separation patterns of the irPCR products obtained from different hybrids this approach allows the analysis and/or isolation of chromosomal sequences in a region-specific manner. In addition, irPCR analysis facilitates combined genetic and physical mapping of a small genomic region (see Chapter 2 and below).

By interrepeat PCR many loci can be simultaneously analysed using only one 'consensus' primer. In a recent study up to 20 genomic polymorphisms were detected as the presence or absence of amplified DNA fragments originating from genomic segments flanked by Alu repeats (Zietkiewicz *et al.*, 1992). The particular primer used directed amplification of DNA segments flanked by Alu repeats in inverted orientation. Simply by using other Alu-specific primers or by manipulating the experimental conditions many more genomic loci can be targeted simultaneously. The resolution of 1D electrophoretic separation will then no longer be sufficient. By 2D electrophoresis hundreds of such fragments can be resolved, greatly increasing the efficiency of a linkage study as described by Zietkiewicz *et al.* (1992). As suggested by Cox and Lehrach (1991) a similar increase in resolution of irPCR products can be obtained by 2D electrophoresis in mouse genome mapping studies based on the use of primers for B1/B2 repetitive sequences or microsatellite motifs.

The major application of 2D DNA typing using the inter-repeat approach is to direct a linkage study to particular genomic regions. This can be accomplished by using irPCR products derived from hybrid cell lines or YAC clones as hybridization probes to target the corresponding loci in a 2D separation pattern of irPCR products from total genomic DNA. In this way polymorphic loci can be detected in a selective manner, i.e. only loci in genomic regions known to contain the gene of interest are screened. This procedure of genomic fine mapping, termed positional 2D DNA typing, provides genetic information that is directly linked to physical map information in the YAC clones used as hybridization probe. Moreover, 2D DNA typing of irPCR products obtained from YACs can be used to order these into contigs on the basis of overlapping spot patterns. Indeed, with the ever-increasing size of YAC inserts (Bellanné-Chantelot

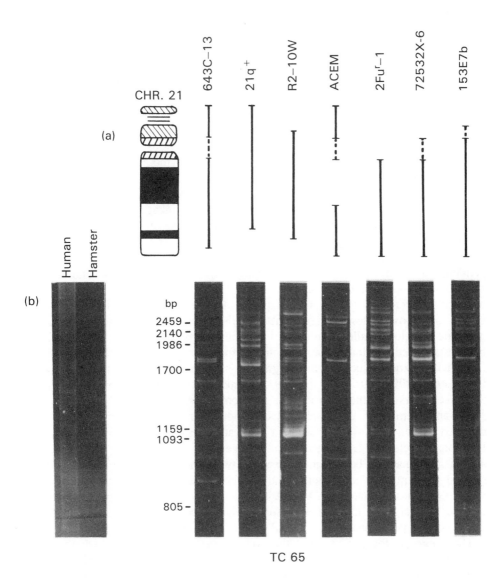

Fig. 5.2. Inter-Alu PCR on a panel of human–hamster somatic cell hybrids containing different parts of human chromosome 21. The primer used was TC65 (Nelson *et al.*, 1989). (a) Schematic depiction of the chromosome 21 content of each of the hybrids used (see also Gardiner *et al.*, 1990; Meulenbelt *et al.*, 1993). (b) The PAAGE results obtained after inter-Alu PCR on genomic DNAs isolated from the hybrids. The left shows inter-Alu PCR products from total human DNA as a positive control and from hamster DNA as a negative control.

et al., 1992) the resolution of one-dimensional techniques may no longer be adequate to determine overlap.

5.3   TWO-DIMENSIONAL cDNA TYPING: ANALYSING THE EXPRESSED PART OF THE GENOME

Current positional cloning technology employs anonymous polymorphic markers and does not allow direct identification of genes through polymorphisms in their coding regions. (An important exception is formed by the category of genetic diseases in which the expansion of a trinucleotide repeat is responsible for the disease phenotype; see also Chapter 2 and below.) We have previously discussed 2D DNA typing based on the analysis of hundreds of highly polymorphic micro- and mini-satellite sequences by using so-called core probes. This 'genome scanning' approach can be applied in analysing total genomes for variation, such as in comparing tumour DNA with normal DNA, and in the linkage analysis of disease pedigrees (see Chapter 4). These studies will result in one or more 2D spots which can be isolated from the gel, PCR amplified, sequenced and mapped to define the position on the chromosome of the genes involved in the disease under study. Depending on the accuracy obtained, this area can be as large as several million basepairs harbouring more than a hundred different genes. Some solution to this problem is offered by genetic fine mapping of the region by positional 2D DNA typing (see above). However, by genetic studies alone it will be very difficult to delineate a region smaller than about 1 million basepairs. Although protocols have been established to clone exons selectively from large pieces of cloned genomic DNA with a known location in the genome such as mapped cosmids and YACs (Duyk *et al.*, 1990; Buckler *et al.*, 1991; Melmer and Buchwald, 1992), to go from the linked marker to the gene itself is far from trivial.

As an alternative, the total mRNA population derived from normal and diseased cells/tissues can be scanned for variations, thereby allowing only the functionally relevant parts (genes) of the genome to be directly analysed. An analytical format can be designed based on two-dimensional separation of cDNA derived from mRNA populations to allow for identification of pathological gene variants. Such a 'functional 2D DNA typing' approach would represent a novel method of analysing, in parallel, all or a selected portion of the coding parts of the genome by 2D analysis of cDNA made from mRNA.

The approach focuses on only the coding part of the genome, i.e. the total population of mRNAs. Coding regions in genes are currently analysed either by Northern blot analysis or from cDNA libraries. Although effective methods to identify and isolate cDNAs for genes that are differentially expressed in various cells or under altered conditions have been developed (see, for example, Liang and Pardee, 1992), it has proved to be more difficult to scan RNA populations for mutations and polymorphisms.

Two-dimensional cDNA typing aims at simultaneously visualizing as many cDNA sequences as possible by separation in a 2D gel and subsequent staining of fragments (by ethidium bromide or silver) and/or hybridization analysis with specific probes. By generating cDNAs (from poly-A$^+$ RNA) using oligo-dT primers attached to a GC-clamp separation conditions are optimized for detecting all possible mutations (Sheffield *et al.*, 1989; Chapter 1). A first prerequisite for a successful application of this principle is the availability of protocols to synthesize full-length cDNAs; otherwise smears will be obtained in the first dimension rather than spots. Such protocols have been reported (Fukuoka and Scheele, 1991). Then, it is important to obtain enough resolving power to separate the about 15 000 mRNAs contained in a typical cell. This problem can be solved by using larger gels. Alternatively, the 2D cDNA separation patterns can be hybridized with probes for sequence motifs (e.g., zinc-finger, homeobox) or with cDNAs selected for position in the genome, for example, through exon trapping. This will allow all or a particular subgroup of functional regions in the genome to be analysed for sequence variation among individuals and different tissues.

A particularly interesting potential application of 2D cDNA typing is the analysis of patients from disease pedigrees such as fragile X, Huntington's chorea and myotonic dystrophy. The defect causing these diseases is related to trinucleotide microsatellite polymorphisms (see, for example, Verkerk *et al.*, 1991). It has been shown that patients suffering from these diseases carry extreme length mutations in [CNG]$_n$ stretches in the disease genes. Furthermore, the phenomenon of genetic anticipation (increasing severity of the disease in patients further down the pedigree) has been shown to correlate with increasing length of the trinucleotide stretch. Using core probes for such trinucleotide motifs, it should be possible to detect directly such mutations in cDNA and/or genomic DNA of patients with these or other unknown genetic diseases.

In general, 2D cDNA typing would allow the simultaneous analysis of polymorphisms in many candidate genes and, hence, could be a useful tool to perform genetic association studies.

5.4   GENE SCANNING: DIAGNOSIS OF ALL POSSIBLE MUTATIONS IN DISEASE GENES

The identification of genes whose malfunction underlies disease phenotypes opens up new routes towards more efficient diagnostics and therapies. The identification of the DNA sequence variation or mutation that makes a normal gene a disease gene is thereby of key importance. Direct detection of the disease-causing gene mutation theoretically allows for 100% accurate diagnosis of individuals being a carrier of the disease or at risk of developing it sooner or later. An overview of some of the characteristics of disease-causing mutations in human disease genes is shown in Table 1.4 (Chapter 1).

The accuracy with which one can determine individuals to be at risk for developing a particular genetic disease or having carrier status is of crucial importance for starting large-scale population screening programs. If this accuracy is too low, the percentage of individuals for whom risk cannot be determined is unacceptably high. The American Society for Human Genetics has stated that large-scale screening, e.g., for cystic fibrosis, can only be done in a responsible way if the accuracy of diagnosis is more than 95%. In general, the accuracy of genetic disease testing can be compromised by two major factors.

(1)   The total spectrum of mutations causing the disease is not fully known. Some genetic diseases are caused by a single basepair mutation at a single site in the gene. An example is the sickle cell mutation in the α-globin gene. In this case screening is 100% accurate, very simple and can be performed nowadays using many different test formats. However, this has been shown to be an exception and not the rule. Many of the newly discovered disease genes are very large and can carry many different types of disease-causing mutations at many different sites in and outside the gene. Examples are the more than 300 different β-thalassaemia mutations in the β-globin gene (Losekoot *et al.*, 1990), the many different mutations (deletions, rearrangements, point mutations) causing Duchenne muscular dystrophy (Bakker, 1989), haemophilia B (Giannelli *et al.*, 1990) and cystic fibrosis (Tsui, 1992). The genes involved in these last three disorders are very large and a still growing number of disease mutations is being described. For the CF gene a frequent mutation has been described (the δF508 3 bp deletion in exon 10) which is the cause of the disease in nearly 70% of the Northern European Caucasian cases. However, more than 300 different mutations have now been described for the CF gene which explain about 2/3 of the other 30% of the mutations. Therefore, in 10% of Caucasian CF cases the disease-causing mutation is still unknown. Furthermore, there can be a population-dependent incidence of these mutations. For example, in Southern Europe the δF508 CF mutation is responsible for only 50% of the cases.

(2)   New and unknown mutations continuously arise. This is related to the previous problem of dealing with large genes. An example of this is the Duchenne muscular dystrophy gene which is the largest gene discovered so far in the human genome with a length of 2.4 million basepairs. Disease-causing mutations have been shown to occur in 1 in 3000 individuals with in more than 30% of these cases new mutations, that is, patients for which there was no previous family history.

These two major compromising factors necessarily limit current testing to only the major disease-causing mutations. In the last five years recombinant DNA research has seen the birth of a wealth of techniques that can rapidly diagnose the presence or absence of a given point mutation. These techniques range from simple assays involving restriction enzyme digestion, oligonucleotide hybridization and minisequencing reactions to elegant assays based on selective PCR amplification or ligation inhibition (see also Chapter 1). Although all of these tests can be used for the detection of a single known point mutation, they are not suitable for comprehensive detection of the spectrum of mutations which underlie the majority of genetic diseases. For this purpose test formats are required that allow the detection of any given variation, also in large genes. The most comprehensive test systems now available are based on the use of allele-specific oligonucleotide probes in hybridization assays with multiplex PCR-amplified genomic DNAs (Shuber *et al.*, 1993) or the inability of PCR primers to function when there is a mismatch at their 3′ OH residue (amplification refractory mutation system; Newton *et al.*, 1989; Ferrie *et al.*, 1992). Although technical improvements will increase the number of mutations that can be detected in one single assay, this technology will never reach accuracies of diagnosis that are close to 100% for diseases such as CF.

Two-dimensional DNA typing has the potential for comprehensive detection of mutations in disease genes. In view of its capacity to resolve large numbers of fragments on the basis of both size and basepair sequence, 2D DNA typing allows even very large genes to be scanned for any given type of mutation ('2D gene scanning'). In this respect it should be noted that DGGE appears to be the most powerful of all available methods in detecting point mutations; nearly all possible basepair substitutions of a given DNA sequence can be detected (Myers *et al.*, 1985b,c; see also Chapter 1).

The power of DGGE in detecting disease-causing mutations has been demonstrated for the CF gene (Fanen *et al.*, 1992). The use of the system in population screening is illustrated in Fig. 5.3a, in which 14 different neutral and pathological sequence variants of exon 10 of the CF gene are shown in 8 different apparently normal individuals (these variants were identified by screening 103 individuals; see also Chapter 1, Fig. 1.5).

On the right part of Fig. 5.3a the segregation of the δF508 mutation is shown in a CF pedigree. In this particular case 3 bands can be observed in the parents. These are derived from homoduplex molecules (the lower two bands) and heteroduplex molecules (the upper bands). The heteroduplex bands are by-products of the PCR (see Chapter 1; Fig. 1.5). The homoduplex bands represent the M470 and V470 alleles. This is a frequently observed neutral sequence variation. The δF508 allele co-migrates with the M470 allele, even in DGGE. The occurrence of neutral mutations together with disease-causing mutations in the same fragment is not exceptional and contributes to the complexity of disease gene diagnosis. In the children only one band can be observed, because they are homozygous for both δF508 and M470.

To distinguish between δF508 and M470 in the children their DNA was heteroduplexed to DNA from an individual lacking both the δF508 and the M470 mutation. In this case individual RH23 was used for this purpose. Now, the presence of δF508 is indicated by the early melting heteroduplex bands (Fig. 5.3a).

Fig. 5.3b shows the same analysis, but now in a 2D format. That is, for each individual the PCR-amplified fragments were first subjected to size separation and subsequently to DGGE. In this case there is no improved resolution other than the presence of spots instead of bands. However, when one needs to analyse all 27 exons of the CFTR gene such a format offers considerable advantages.

This is illustrated in Fig. 5.3c in which multiplex PCR products corresponding to 16 exons of the CFTR gene were separated by 2D electrophoresis. In several of the exons mutations could be detected showing that many different sequence variants can be detected in a uniform format.

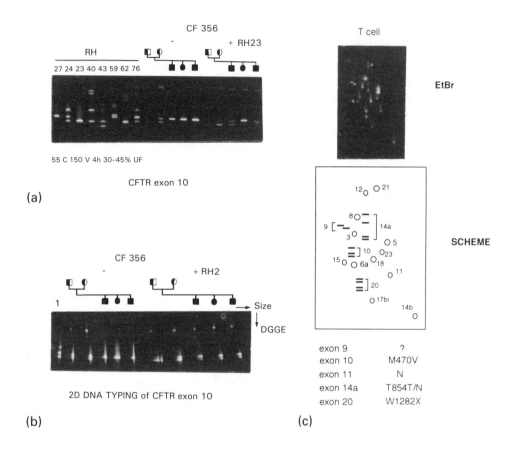

Fig. 5.3. (a) Screening of different normal and CF individuals by DGGE analysis. On the left different DGGE patterns observed among a sample of 103 different individuals are shown. On the right a pedigree is shown with (+) and without (−) adding a genomic DNA (RH23) homozygous for the V470 mutation to the PCR amplification mixture. This allows the δF508 mutation to be detected in sibs homozygous for the δF508 mutation and the M470 variant. (b) Two-dimensional separation of the sequence variants from the CF pedigree shown in (a). Each size-separated 336 bp PCR product was loaded on a second dimension DGGE gel, resulting in sharply focused spots. Individual 1 on the left is heterozygous M470/V470. (c) 2D separation of multiplex PCR products obtained from a normal individual (T cell) using GC-clamped primers for 16 exons of the CFTR gene. The primers have been described by Fanen et al. (1992). On top the EtBr stained 2D gel separation pattern is shown while below a schematic representation is shown. In the scheme empty dots are homozygous normal sequences while black strips indicate sequence variants; numbers refer to the CFTR exon number. Below the scheme the different (neutral sequence variations observed in this individual are indicated for each exon. W1282X is a CF mutation.

Generally, in 2D gene scanning the gene of interest is first divided in fragments which can be generated by means of PCR amplification. These fragments can cover selected areas, such as the coding region, the RNA splice sites and the 5′ and 3′ non-coding adjacent regions or simply comprise the complete gene sequence. For all of these fragments the oligonucleotides are chosen in such a way that optimal separation in both the first dimension as well as the DGGE dimension can be accomplished. This will include the addition of a 'GC-clamp' sequence to one of the PCR primers and heteroduplex analysis. For the fragments individually and as a group optimal DGGE separation formats are determined, empirically and by using the computer melting program (see Chapter 1). Whereas wild-type fragments will be present as a single spot, mutant variants will result in the detection of additional spots derived from the heteroduplex fragments (see Fig. 5.3c).

An alternative format for 2D gene scanning can include hybridization analysis using cDNA probes for the genes of interest, rather than PCR analysis. This approach offers the advantages of being both cheaper (no need for many expensive primers) and simpler (only one hybridization reaction). This is especially true when one aims at analysing the gene in its entirety, including introns and flanking (regulatory) regions. A disadvantage, however, is that no GC-clamping or heteroduplexing is possible, which reduces the scanning efficiency of DGGE by about 60% (see Chapter 1; Table 1.2).

5.5 CONCLUDING REMARKS

The concept of parallel screening of large numbers of DNA fragments for finding unknown genes or to perform accurate molecular diagnosis offers a possible solution to the problem of heterogeneity. Heterogeneity in DNA sequence information is fundamental to all phenomena in the living world: the creation of a bewildering number of possible phenotypes. This can vary from disease phenotypes (in humans, animals and plants) to economically important traits (e.g. milk production in cattle, disease resistance in plants). Thus far this heterogeneity has been addressed in a serial manner, that is, the most interesting DNA sequence variants were singled out and studied one by one. The success of this approach has led to an ever-higher demand for methods and procedures to increase efficiency. In other words, what used to be a haphazard approach has become a fully institutionalized strategy to go from individual or group differences in phenotypes to the DNA or RNA variants underlying the phenomenon of interest.

Based on 2D DNA typing as a method for total genome scanning using VNTR sequence loci as anchor points, a number of additional applications of the general '2D principle' have been anticipated in this final chapter. It is important to realize that none of these potential applications has yet been brought into practice. We simply selected several of the, in our opinion, most promising spin-offs of the basic technology as described in the previous chapters.

References

Aaltonen, L.A., Peltomäki, P., Leach, F.S. Sistonen, P., Pylkkanen, L., Mecklin, J.-P., Jarvinen, H., Powell, S.M., Jen, J., Hamilton, S.R., Petersen, G.M., Kinzler, K.W., Vogelstein, B., and de la Chapelle, A. (1993) Clues to the pathogenesis of familial colorectal cancer. *Science* **260,** 812–816.

Abrams, E.S., Murdaugh, S.E., and Lerman, L.S. (1990) Comprehensive detection of single base changes in human genomic DNA using denaturing gradient gel electrophoresis and a GC clamp. *Genomics* **7,** 463–475.

Ainsworth, P.J., Surh, L.C., and Coulter-Mackie, M.B. (1991) Diagnostic single strand conformational polymorphism (SSCP): a simplified non-radioisotopic method as applied to a Tay-Sachs B1 variant. *Nucleic Acids Res.* **19,** 405.

Akli, S., Chelly, J., Lacorte, J.-M., Poenaru, L., and Kahn, A. (1991) Seven novel Tay–Sachs mutations detected by chemical mismatch cleavage of PCR-amplified cDNA fragments. *Genomics* **11,** 124–134.

Ali, I.U., Campbell, G., Merlo, G.R., Smith, G.H., Callahan, R., and Lidereau, R. (1989) Multiple genetic alterations in human breast cancer and their possible prognostic significance. In: Furth, M., and Greaves, M. (eds), *Cancer Cells. 7.* Molecular Diagnostics of Human Cancer, Cold Spring Harbor Laboratory Press, pp. 399–404.

Ali, S., and Wallace, B. (1989) Enzymatic synthesis of DNA probes complementary to a variable number tandem repeat locus. *Anal. Biochem.* **179,** 280–283.

Ali, S., Müller, C.R., and Epplen, J.T. (1986) DNA fingerprinting by oligonucleotide probes specific for simple repeats. *Hum.Genet.* **74,** 239–243.

Anderson, N.L. (1991) *Two-dimensional Electrophoresis: Operation of the ISO-DALT System.* 2nd edn, Large Scale Biology Press, Rockville, MD.

Anderson, N.L., Taylor, J., Scandora, A.E., Coulter, B.P., and Anderson, N.G. (1981) The Tycho system for computer analysis of two-dimensional gel electrophoresis patterns. *Clin. Chem.* **27,** 1807–1820.

Anderson, S., Bankier, A.T., Barrell, B.G., deBruyn, M.H.L., Coulson, A.R., Droin, J., Eperon, I.C., Nierlich, D.P., Roe, B.A., Sanger, F., Schrier, P.H., Smith, A.J.H., Staden, R., and Young, I.G. (1981) Sequence and organization of the human mitochondrial genome. *Nature* **290,** 457–461.

Armour, J.A.L., Patel, I., Thein, S.L., Fey, M.F., and Jeffreys, A.J. (1989) Analysis of somatic mutations at human minisatellite loci in tumors and cell lines. *Genomics,* **4,** 328–334.

Armour, J.A.L., Povey, S., Jeremiah, S., and Jeffreys, A.J. (1990) Systematic cloning of human minisatellites from ordered array charomid libraries. *Genomics* **8,** 501– 512.

Attree, O., Vidaud, D., Amselem, S., Lavergne, J.-M., and Goossens, M. (1989) Mutations in the catalytic domain of human coagulation factor IX: rapid characterization by direct genomic sequencing of DNA fragments displaying an altered melting behaviour. *Genomics* **4,** 266–272.

Au, L.-C., Ts'o, P.O.P., and Yi, M. (1990) Monitoring mammalian genome rearrangements with mid-repetitive sequences as probes. *Anal. Biochem.* **190,** 326–330.

Baker, S.J., Fearon, E.R., Nigro, J.M., Hamilton, S.R., Preisinger, A.C., Jessup, J.M., vanTuinen, P., Ledbetter, D.H., Barker, D.F., Nakamura, Y., White, R., and Vogelstein, B. (1989) Chromosome 17 deletions and p53 gene mutations in colorectal carcinoma. *Science* **244,** 217–221.

Baker, S.J., Preisinger, A.C., Jessup, J.M., Paraskeva, C., Markowitz, S., Wilson, J.K.V., Hamilton, S., and Vogelstein, B. (1990) p53 gene mutations occur in combination with 17p allelic deletions as late events in colorectal tumorigenesis. *Cancer Res.* **50,** 7717–7722.

Bakker, E. (1989) Duchenne muscular dystrophy — carrier detection and prenatal diagnosis by DNA analysis — new mutation and mosaicism. *Thesis.* Leiden.

Barker, D., Schafer, M., and White, R. (1984) Restriction sites containing CpG show a higher frequency of polymorphism in human DNA. *Cell* **36,** 131–138.

Bautsch, W. (1988) Rapid physical mapping of the *Mycoplasma mobile* genome by two-dimensional field inversion gel electrophoresis. *Nucleic Acids Res.* **16** (24), 11461–11467.

Beckmann, J.S., and Weber, J.L. (1992) Survey of human and rat microsatellites. *Genomics* **12,** 627–631.

van Belkum, A., Ramesar, J., Trommelen, G.J.J.M., and Uitterlinden, A.G. (1992) Mini- and micro-satellites in the genome of rodent malaria parasites. *Gene* **118,** 81–86.

Bell, G.I., Selby, M.J., and Rutter, W.J. (1982) The highly polymorphic region near the human insulin gene is composed of simple tandemly repeating sequences. *Nature* **295,** 31–35.

Bell, L., and Byers, B. (1983) Separation of branched from linear DNA by two-dimensional gel electrophoresis. *Anal. Biochem.* **130,** 527–535.

Bellanné-Chantelot, C., Lacroix, B., Ougen, P., Billault, A., Beaufils, S., Bertrand, S., Georges, I., Gilbert, F., Gros, I., Lucotte, G., Susini, L., Codani, J.-J., Gesnouin, P., Pook, S., Vaysseix, G., Lu-kuo, J., Ried, T., Ward, D., Chumakov, I., le Paslier, D., Barillot, E., and Cohen, D. (1992) Mapping the whole human genome by fingerprinting yeast artificial chromosomes. *Cell* **70,** 1059–1068.

Bewsey, K.E., Johnson, M.E., and Huff, J.P. (1991) A method for the rapid detection of single base polymorphisms in genomic DNA. *Biotechniques* **10,** 725.

Blanchetot, A. (1991) A *Musca domestica* satellite sequence detects individual polymorphic regions in insect genome. *Nucleic Acids Res.* **19**, 929–932

Boehm, T.L.J., and Drahovsky, D. (1984) Two-dimensional restriction mapping by digestion with restriction endonucleases of DNA in agarose and polyacrylamide gels. *J. Biochem. Biophys. Methods* **9**, 153–161.

Børresen, A.-L., Hovig, E., and Brogger, A. (1988) Detection of base mutations in genomic DNA using denaturing gradient gel electrophoresis (DGGE) followed by transfer and hybridization with gene-specific probes. *Mutat. Res.* **202**, 77–83.

Børresen, A.-L., Hovig, E., Smith-Sorensen, B., Malkin, D., Lystad, Andersen, T.I., Nesland, J.M., Isselbacher, K.J., and Friend, S.H. (1991) Constant denaturant gel electrophoresis as a rapid screening technique for p53 mutations. *Proc. Natl. Acad. Sci. USA* **88**, 8405–8409.

Botstein, D., White, R., Skolnick, M., and Davis, R. (1980) Construction of a genetic linkage map in man using restriction fragment length polymorphisms. *Am. J. Hum. Genet.* **32**, 314–331.

Boylan, K.B., Ayres, T.M., Popko, B., Takahashi, N., Hood, L., and Prusiner, S.B. (1990) Repetitive DNA (TGGA)$_n$ to the human myelin basic protein gene: a new form of oligonucleotide repetitive sequence showing length polymorphism. *Genomics* **6**, 16–22.

Branda, R.F., Sullivan, L.M., O'Neill, J.P., Falta, M.T., Nicklas, J.A., Hirsch, B., Vacek, P.M., and Albertini, R.J. (1993) Measurement of HPRT mutant frequencies in T-lymphocytes from healthy human populations. *Mutation Res.* **285**, 267–279.

Brilliant, M.H., Gondo, Y., and Eicher, E.M. (1991) Direct molecular identification of the mouse pink-eyed unstable mutation by genome scanning. *Science* **252**, 566–569.

Britten, R.J., and Kohne, D.E. (1968) Repeated sequences in DNA. *Science* **161**, 529–540.

Brooks-Wilson, A.R., Goodfellow, P.N., Povey, S., Nevanlinna, H.A., de Jong, P.J., and Goodfellow, P.J. (1990) Rapid cloning and characterization of new chromosome 10 DNA markers by Alu element mediated PCR. *Genomics* **7**, 614–620.

Buckler, A.J., Chang, D.D., Graw, S.L., Brook, J.D., Haber, D.A., Sharp, P.A., and Housman, D.E. (1991) Exon amplification: a strategy to isolate mammalian genes based on RNA splicing. *Proc. Natl. Acad. Sci. USA* **88**, 4005–4009.

Burghes, A.H.M., Logan, C., Hu, X., Bellfall, B., Worton, R.G., and Ray, P.N. (1987) A cDNA clone from the Duchenne/Becker muscular dystrophy gene. *Nature* **328**, 434–437.

Burke, D.T., Carle, G.F., and Olson, M.V. (1987) Cloning of large segments of DNA into yeast by yeast artificial chromosome vectors. *Science* **236**, 806–812.

Burmeister, M., and Lehrach, H. (1986) Longe-range restriction map around the Duchenne muscular dystrophy gene. *Nature* **324**, 482–485.

Burmeister, M., diSibio, G., Cox, R., and Myers, R.M. (1991) Identification of polymorphisms by genomic denaturing gradient gel electrophoresis: application to the proximal region of human chromosome 21. *Nucleic Acids Res.* **19**, 1475–1481.

Buroker, N., Bestwick, R., Haight, G., Magenis, R.E., and Litt, M. (1987) A hypervariable repeated sequence on human chromosome 1p36. *Hum. Genet.* **77**, 175–181.

Burton, F.H., Loeb, D.D., Chao, S.F., Hutchinson III, C.A., and Edgell, M.H. (1985) Transposition of a long member of the L1 major interspersed DNA family into the mouse beta globin gene locus. *Nucleic Acids Res.* **13**, 5071–5084.

Caetano-Anollés, G., Bassam, B.J., and Gresshoff, P.M. (1991) DNA amplification fingerprinting using very short arbitrary oligonucleotide primers. *Bio/technology* **9**, 553–556

Cantor, C.R. (1990) Orchestrating the Human Genome Project. *Science,* **248**, 49–51.

Capon, D.J., Chen, E.Y., Levinson, A.D., Seeburg, P.H., and Goeddel, D.V. (1983) Complete nucleotide sequences of the T24 human bladder carcinoma oncogene and its normal homologue. *Nature,* **302**, 33–37.

Cariello, N.F., Scott, J.K., Kat, A.G., and Thilly, W.G. (1988a) Resolution of a missense mutant in human genomic DNA by denaturing gradient gel electrophoresis and direct sequencing using *in vitro* DNA amplification: HPRT$_{\text{Munich}}$ *Am. J, Hum. Genet.* **42**, 726–734.

Cariello, N.F., Keohavong, P., Sanderson, B.J.S., and Thilly, W.G. (1988b) DNA damage produced by ethidium bromide staining and exposure to ultraviolet light. *Nucleic Acids Res.* **16**, 4157.

Cariello, N.F., Swenberg, J. A., De Bellis, A., and Skopek, T. R. (1991a) Analysis of mutations using PCR and denaturing gradient gel electrophoresis. *Environ. Mol. Mutagen.* **18**, 249–254.

Cariello, N.F., Swenberg, J.A., and Skopek, T.R. (1991b) Fidelity of Thermococcus litoralis DNA polymerase (Vent^{TM}) in PCR determined by denaturing gradient gel electrophoresis. *Nucleic Acids Res.* **19**, 4193–4198.

Carle, G.F., and Olson, M.V. (1984) Separation of chromosomal DNA molecules from yeast by orthogonal-field-alternation gel electrophoresis. *Nucleic Acids Res.* **12**, 5647–5664.

Carle, G.F., and Olson, M.V. (1985) An electrophoretic karyotype for yeast. *Proc. Natl. Acad. Sci. USA* **82**, 3756–3760.

Carle, G.R., Frank, M., and Olson, M.V. (1986) Electrophoretic separations of large DNA molecules by periodic inversion of the electric field. *Science* **232**, 65–68.

Carothers, A.M., Urlaub, G., Grunberger, D., and Chasin, L.A. (1988) Mapping and characterization of mutations induced by benzo[*a*] pyrene diol epoxide at dihydrofolate reductase locus in CHO cells. *Som. Cell Mol. Genet.* **14**, 169–183.

Carrano, A.V., Lamerdin, J., Ashworth, L.K., Watkins, B., Branscomb, E., Slezak, T., Raff, M., De Jong, P.J., Keith, D., McBride, L., Meister, S., and Kronick, M. (1989) A high-resolution, fluorescence-based, semiautomated method for DNA-fingerprinting. *Genomics* **4**, 129–136.

Casna, N.J., Novack, D.F., Hsu, M.-T., and Ford, J. (1986) Genomic aalysis II: isolation of high molecular weight heteroduplex DNA following differential methylase protection and Formamide-PERT hybridization. *Nucleic Acids Res.* **14**, 7285–7303.

Cavenee, W.K. (1989) Loss of heterozygosity in stages of malignancy. *Clin. Chem.* **35**, 7B, B48–B52.

Cawthon, R.M., Weiss, R., Xu, G., Viskochil, D., Culver, M., Stevens, J., Robertson, M., Dunn, D., Gesteland, R., O'Connell, P., and White, R. (1990) A major segment of the neurofibromatosis type 1 gene: cDNA sequence, genomic structure, and point mutations. *Cell* **62,** 193–201.

Chamberlain, J.S., Boehnke, M., Frank, T.S., Kiousis, S., Xu, J., Guo, S.-W., Hauser, E.R., Norum, R.A., Helmbold, E.A., Markel, D.S., Keshavarzi, S.M., Jackson, C.E., Calzone, K., Garber, J., Collins, F.S., and Weber, B.L. (1993) BRCA1 maps proximal to D17S579 on chromosome 17q21 by genetic analysis. *Am. J. Hum. Genet.* **52,** 792–798.

Chang, J.C., and Kan, Y.W. (1982) A sensitive new prenatal test for sickle cell anemia. *N. Engl. J. Med.* **307,** 30–32.

Chebloune, Y., Trabuchet, G., Poncet, D., Cohen-Solal, M., Faure, C., Verdier, G., and Nigon, V.M. (1984) A new method for detection of small modifications in genomic DNA, applied to the human δ-β globin gene cluster. *Eur. J. Biochem.* **142,** 473–480.

Chumakov, I., Rigault, P., Guillou, S., Ougen, P., Billaut, A., Guasconi, G., Gervy, P., leGall, I., Soularue, P., Grinas, L., Bougueleret, L., Bellanné-Chantelot, C., Lacroix, B., Barillot, E., Gesnouin, P., Pook, S., Vaysseix, G., Frelat, G., Schmitz, A., Sambucy, J.-L., Bosch, A., Estivill, X., Weissenbach, J., Vignal, A., Riethman, H., Cox, D., Patterson, D., Gardiner, K., Hattori, M., Sakaki, Y., Ichikawa, H., Ohki, M., le Paslier, D., Heilig, R., Antonarakis, S., and Cohen, D. (1992) Continuum of overlapping clones spanning the entire human chromosome 21q. *Nature* **359,** 380–387.

Church, G.M., and Gilbert, W. (1984) Genomic sequencing. *Proc. Natl. Acad. Sci. USA* **81,** 1991–1995.

Clay, T.M., Bidwell, J.L., Howard, M.R., and Bradley, B.A. (1991) PCR-fingerprinting for selection of HLA matched unrelated marrow donors. *Lancet* **337,** 1049–1052.

Clerget-Darpoux, F., and Bonaiti-Pellié, C. (1992) Strategies based on marker information for the study of human diseases. *Ann. Hum. Genet.* **56,** 145–153.

Collins, F.S. (1992) Positional cloning: Let's not call it reverse anymore. *Nature Genet.* **1,** 3–6.

Collins, M., and Myers, R.M. (1987) Alterations in DNA helix stability due to base modifications can be evaluated using denaturing gradient gel electrophoresis. *J. Mol. Biol.* **198,** 737–744.

Conner, B.J., Reyes, A.A., Morin, C., Itakura, K., Teplitz, R.L., and Wallace, R.B. (1983) Detection of sickle cell β-globin allele by hybridization with synthetic oligonucleotides. *Proc. Natl. Acad. Sci. USA* **80,** 278–282.

Cooper, D.N., Smith, B.A., Cooke, H., Niemann, S., and Schmidtke, J. (1985) An estimate of unique DNA sequence heterozygosity in the human genome. *Hum. Genet.* **69,** 201–205.

Costes, B., Girodon, E., Ghanem, N., Chassignol, M., Thuong, N.T., Dupret, P., and Goossens, M. (1993) Psoralen-modified oligonucleotide primers improve detection of mutations by denaturing gradient gel electrophoresis and provide an alternative to GC-glamping. *Hum. Mol. Genet.* **2,** 393–397.

Cotton, R.G.H. (1993) Current methods of mutation detection. *Mutat. Res.* **285,** 125–144.

Cotton, R.G.H., and Campbell, R.D. (1989) Chemical reactivity of matched cytosine and thymine bases near mismatched and unmatched bases in a heteroduplex between DNA strands with multiple differences. *Nucleic Acids Res.* **17,** 4223–4233.

Cotton, R.G.H., Rodrigues, N.R., and Campbell, R.D. (1988) Reactivity of cytosine and thymine in single-basepair mismatches with hydroxylamine and osmium tetroxide and its application to the study of mutations. *Proc. Natl. Acad. Sci. USA* **85,** 4397–4401.

Coulson, A., and Sulston, J. (1988) Genome mapping by restriction fingerprinting. In: K.E. Davies (ed.), *Genome Analysis: A Practical Approach,* IRL Press, Oxford, pp. 19–40.

Cox, D. (1991) Radiation hybrids. *Science* **234,** 61–65

Cox, R.D., and Lehrach, H. (1991) Genome mapping: PCR based meiotic and somatic cell hybrid analysis. *BioEssays* **13,** 193–198.

Cox, R.D., Copeland, N.G., Jenkins, N.A., and Lehrach, H. (1991) Interspersed repetitive element polymerase chain reaction product mapping using a mouse interspecific backcross. *Genomics* **10,** 375–384.

Craig, A.G., Nizetic, D., Hoheisel, J.D., Zehetner, G., and Lehrach, H. (1990) Ordering of cosmid clones covering the herpes simplex virus type 1 (HSV-1) genome: a test case for fingerprinting by hybridisation. *Nucleic Acids Res.* **18,** 2653–2660.

Dallas, J.F. (1988) Detection of DNA 'fingerprints' of cultivated rice by hybridization with a human minisatellite probe. *Proc. Natl. Acad. Sci. USA* **85,** 6831–6835.

Danenberg, P.V., Horikoshi, T., Volkenandt, M., Danenberg, K., Lenz, H.-J., Shea, L.C.C., Dicker, A.P., Simoneau, A., Jones, P.A., and Bertino, J.R. (1992) Detection of point mutations in human DNA by analysis of RNA conformation polymorphism(s). *Nucleic Acids Res.* **20,** 573–579.

Dausset, J., Cann, H., Cohen, D., Lathrop, M., Lalouel, J.-M., and White, R. (1990) Centre d'Etude Polymorphisme Humain (CEPH): collaborative genetic mapping of the human genome. *Genomics* **6,** 575–577.

Davis, L.M., Fairfield, F.R., Harger, C.A., Jett, J.H., Keller, R.A., Hahn, J.H., Krakowski, L.A., Marrone, B.L., Martin, J.C., Nutter, H.L., Ratliff, R.L., Brooks Shera, E., Simpson, D.J., and Soper, S.A. (1991) Rapid DNA sequencing based upon single molecule detection. *Genet. Anal. Technol. Appl.* **8,** 1–7.

Dean, M., White, M., Amos, J., Gerrard, B., Stewart, C., Khaw, K.-T., and Leppert, M. (1990) Multiple mutations in highly conserved residues are found in mildly affected cystic fibrosis patients. *Cell* **61,** 863–870.

Delattre, O., Olschwang, S., Law, D.J., Melot, T., Remvikos, Y., Salmon, R.J., Sastre, X., Validire, P., Feinberg, A.P., and Thomas, G. (1989) Multiple genetic alterations distinguish distal from proximal colorectal cancer. *Lancet* Vol. 2, 353–356.

Delwart, E. L., Sphaer, E. G., Louwagie, J., McCutchan, F.E., Grez, M., Rübsamen-Waigmann, H., and Mullins, J.I. (1993) Genetic relationships determined by a DNA heteroduplex assay: analysis of HIV-1 *env* genes. *Science* **262,** 1257–1261.

Demers, D.B., Odelberg, S.J., and Fisher, L.M. (1990) Identification of a factor IX point mutation using SSCP analysis and direct sequencing. *Nucleic Acids Res.* **18**, 5575.

Devilee, P., Kievits, T., Waye, J.S., Pearson, P.L., and Willard, H.F. (1988) Chromosome-specific alpha satellite DNA: isolation and mapping of a polymorphic alphoid repeat from human chromosome 10. *Genomics* **3**, 1–7.

Devilee, P., Van den Broek, M., Kuipers-Dijkshoorn, N., Kolluri, R., Meera Khan, P., Pearson, P.L., and Cornelisse, C.J. (1989) At least four different chromosomal regions are involved in loss of heterozygosity in human breast carcinoma. *Genomics* **5**, 554–560.

Devilee, P., Cornelisse, C.J., Kuipers-Dijkshoorn, N., Jonker, C., and Pearson, P.L. (1990) Loss of heterozygosity on 17p in human breast carcinomas: defining the smallest common region of deletion. *Cytogenet. Cell Genet.* **53**, 52–54.

Dianzani, I., Camaschella, C., Saglio, G., Forrest, S.M., Ramus, S., and Cotton, R.G.H. (1991a) Simultaneous screening for β-thalassemia mutations by chemical cleavage of mismatch. *Genomics* **11**, 48–53.

Dianzani, I., Forrest, S.M., Camaschella, C., Saglio, G., Ponzone, A., and Cotton, R.G.H. (1991b) Screening for mutations in the phenylalanine hydroxylase gene from Italian patients with phenylketonuria by using the chemical cleavage method: a new splice mutation. *Am J. Hum. Genet.* **48**, 631–635.

Dickson, L.A., Pihlajaniemi, P., Deak, S., Pope, F.M., Nichols, A., Prockop, D.J., and Myers, J.C. (1984) Nuclease S_1 mapping of a homozygous mutation in the carboxylpropeptide-coding region of the proα2(I)collagen gene in a patient with osteogenesis imperfecta. *Proc. Natl. Acad. Sci. USA* **81**, 4524–4528.

Dietrich, W., Katz, H., Lincoln, S.E., Shin, H.-S., Friedman, J., Dracopoli, N., and Lander, E. (1992) A genetic map of the mouse suitable for typing intraspecific crosses. *Genetics* **131**, 423–447.

Dlouhy, S.R., Schaff, D.A., Trofatter, J.A., Liu, H.-S., Stambrook, P., and Tischfield, J.A. (1989) Denaturing gradient gel analysis of single-base substitutions at a mouse adenine phosphoribosyl transferase splice acceptor site. *Mol. Carc.* **2**, 217–225.

Dombroski, B.A., Kazazian, H.H., and Scott, A.F. (1989) Isolation and characterization of a putative functional L1 retroposon. *Am. J. Hum. Genet.* **45**, A183.

Donis-Keller, H., Green, P., Helms, C. Cartinhour, S., Weiffenbach, B., Stephens, K., Keith, T.P., Bowden, D.W., Smith, D.R., Lander, E.S., Botstein, D., Akots, G., Rediker, K.S., Gravius, T., Brown, V.A., Rising, M.B., Parker, C., Powers, J.A., Watt, D.E., Kauffman, E.R., Bricker, A., Phipps, P., Muller-Kahle, H., Fulton, T.R., Ng, S., Schumm, J.W., Braman, J.C., Knowlton, R.G., Barker, D.F., Crooks, S.M., Lincoln, S.E., Daly, M.J., and Abrahamson, J. (1987) A genetic linkage map of the human genome. *Cell* **51**, 319–337.

Drmanac, R., Labat, I., Brukner, I., and Crkvenjakov, R. (1989) Sequencing of megabase plus DNA by hybridisation: theory of the method. *Genomics* **4**, 114–128.

Drmanac, R., Drmanac, S., Strezoska, Z., Paunesku, T., Labat, I., Zeremski, M., Snoddy, J., Funkhouser, W.K., Koop, B., Hood, L., and Crkvenjakov, R. (1993) DNA sequence determination by hybridization: a strategy for efficient large-scale sequencing. *Science* **260**, 1649–1652.

Dryja, T.P., Hahn, L.B., Cowley, G.S., McGee, T.L., and Berson, E.L. (1991) Mutation spectrum of the rhodopsin gene among patients with autosomal dominant retinis pigmentosa. *Proc. Natl. Acad. Sci. USA* **88**, 9370–9374.

Dubrova, Y.E., Jeffreys, A.J., and Malashenko, A.M. (1993) Mouse minisatellite mutations induced by ionizing radiation. *Nature Gene.* **5**, 92–94.

Duyk, G.M., Kim, S., Myers, R.M., and Cox, D.R. (1990) Exon trapping: a genetic screen to identify candidate transcribed sequences in cloned mammalian genomic DNA. *Proc. Natl. Acad. Sci. USA* **87**, 8995–8999.

Economou, E.P, Bergen, A.W., Warren, A.C., and Antonarakis, S.E. (1990) The polydeoxyadenylate tract of Alu repetitive elements is polymorphic in the human genome. *Proc. Natl. Acad. Sci. USA* **87**, 2951–2954.

Economou-Pachnis, A., and Tsichlis, P.N. (1985) Insertion of an Alu SINE in the human homologue of the Mlvi-2 locus. *Nucleic Acids Res.* **13**, 8379–8387.

Economou-Pachnis, A., Lohse, M.A., Furano, A.V., and Tsichlis, P.N. (1985) Insertion of long interspersed repeated elements at the *Igh* (immunoglobulin heavy chain) and *Mlvi-2* (Moloney leukemia virus integration 2) loci of rats. *Proc. Natl. Acad. Sci. USA* **82**, 2857–2861.

Edwards, A., Voss, H., Rice, P., Civitello, A., Stegemann, J., Schwager, C., Zimmermann, J., Erfle, H., Caskey, C.T., and Ansorge, W. (1990) Automated DNA sequencing of the human HPRT locus. *Genomics* **6**, 593–608.

Edwards, A., Civitello, A., Hammond, H.A., and Caskey, C.T. (1991) DNA typing and genetic mapping with trimeric and tetrameric repeats. *Am. J. Hum. Genet.* **49**, 746–756.

Edwards, A., Hammond, H.A., Jin, L., Caskey, C.T., and Chakraborty, R. (1992) Genetic variation at five trimeric and tetrameric repeat loci in four human population groups. *Genomics* **12**, 241–253.

Emi, M., Hata, A., Robertson, M., Iverius, P.-H., Hegele, R., and Lalouel, J.-M. (1990) Lipoprotein lipase deficiency resulting from a nonsense mutation in exon 3 of the lipoprotein lipase gene. *Am. J. Hum. Genet.* **47**, 107–111.

Epstein, N., Nahor, O., and Silver, J. (1990) The 3′ ends of Alu repeats are highly polymorphic. *Nucleic Acids Res.* **18**, 4634.

Fanen, P., Ghanem, N., Vidaud, M., Besmond, C., Martin, J., Costes, B., Plassa, F., and Goossens, M. (1992) Molecular characterization of cystic fibrosis: 16 novel mutations identified by analysis of the whole cystic fibrosis conductance transmembrane regulator (CFTR) coding regions and splice site junctions. *Genomics* **13**, 770–776.

Fanning, T.G., Hu, W.-S., and Cardiff, R.D. (1985) Analysis of tissue-specific methylation patterns of mouse mammary tumor virus by two-dimensional Southern blotting. *J. Virol.* **54** (3), 726–730.

Fearon, E.R., and Vogelstein, B. (1990) A genetic model for colorectal tumorigenesis. *Cell* **61**, 759–767.

Feinberg, A.P., and Vogelstein, B. (1983) A technique for radiolabelling DNA restriction endonuclease fragments to high specific activity. *Anal. Biochem.* **132**, 6–13.

Feinberg, A.P., and Vogelstein, B. (1984) A technique for radiolabelling DNA restriction endonuclease fragments to high specific activity. *Anal. Biochem. Addendum* **137**, 266–267.

Ferrie, R.M., Schwarz, M.J., Robertson, N.H., Vaudin, S., Super, M., Malone, G., and Little, S. (1992) Development, multiplexing, and application of ARMS tests for common mutations in the CFTR gene. *Am. J. Hum. Genet.* **51**, 251–262.

Finkelstein, J.E., Francomano, C.A., Brusilow, S.W., and Traystman, M.D. (1990) Use of denaturing gradient gel electrophoresis for detection of mutation and prospective diagnosis in late onset ornithine transcarbamoylase deficiency. *Genomics* **7**, 167–172.

Fischer, S.G., and Lerman, L.S. (1979a) Length–independent separation of DNA restriction fragments in two–dimensional gel electrophoresis. *Cell* **16**, 191–200.

Fischer, S.G., and Lerman, L.S. (1979b) Two–dimensional electrophoretic separation of restriction enzyme fragments of DNA. *Methods Enzymol.* **68**, 183–191.

Fischer, S.G., and Lerman, L.S. (1980) Separation of random fragments of DNA according to properties of their sequences. *Proc. Natl. Acad. Sci. USA* **77**, 4420–4424.

Fischer, S.G., and Lerman, L.S. (1983) DNA fragments differing by single basepair substitutions are separated in denaturing gradient gels: correspondence with melting theory. *Proc. Natl. Acad. Sci. USA* **80**, 1579–1583.

Flavell, R.A., Kooter, J.M., De Boer, E., Little, P.F.R., and Williamson, R. (1978) Analysis of the β-δ-globin gene loci in normal and Hb Lepore DNA: Direct determination of gene linkage and intergene distance. *Cell* **15**, 25–41.

Forrest, S.M., Dahl, H.-H., Howells, D.W., Dianzani, I., and Cotton, R.G.H. (1991) Mutation detection in phenylketonuria using the chemical cleavage of mismatch method: importance of using probes from both normal and patient samples. *Am. J. Hum. Genet.* **49**, 175–183.

Fu, Y.-H., Kuhl, D.P.A., Pizutti, A., Pieretti, M., Sutcliffe, J.S., Richards, S., Verkerk, A.J.M.H., Holden, J.J.A., Fenwick, R.G., Warren, S.T., Oostra, B.A., Nelson, D.L., and Caskey, C.T. (1991) Variation of the CGG repeat at the fragile X site results in genetic instability: resolution of the Sherman paradox. *Cell* **67**, 1047–1058.

Fu, Y.-H., Pizutti, A., Fenwick, R.G., King, J., Rajnarayan, S., Dunne, P.W., Dubbel, J., Nasser, G.A., Ashizawa, T., de Jong, P.J., Wieringa, B., Korneluk, R., Perryman, M.B., Epstein, H.F., and Caskey, C.T. (1992) An unstable triplet repeat in a gene related to myotonic muscular dystrophy. *Science* **255**, 1256–1258.

Fukuoka, S., and Scheele, G.A. (1991) Novel strategy for synthesis of full-length double-stranded cDNA transcripts without dC-dG tails. *Nucleic Acids Res.* **19**, 6961–6962.

Galinski, M.R., Arnot, D.E., Cochrane, A.H., Barnwell, J.W., Nussenzweig, R.S., and Enea, V. (1987) The circumsporozoite gene of the *Plasmodium cynomolgi* complex. *Cell* **48**, 311–319.

Ganguly, A., and Prockop, D.J. (1990) Detection of single-base mutations by reaction of DNA heteroduplexes with a water-soluble carbodiimide followed by primer extension: application to products from the polymerase chain reaction. *Nucleic Acids Res.* **18**, 3933–3939.

Ganguly, A., Rooney, J.E., Hosomi, S., Zeiger, A.R., and Prockop, D.J. (1989) Detection and location of single-base mutations in large DNA fragments by immunomicroscopy. *Genomics* **4**, 530–538.

Gardiner, K., Horisberger, M., Kraus, J., Tantravahi, U., Korenberg, J., Rao, V., Reddy, S., and Patterson, D. (1990) Analysis of human chromosome 21: correlation of physical and cytogenetic maps; gene and CpG island distributions. *EMBO J.* **9**, 25–34.

Garrels, J.I. (1989) The QUEST system for quantitative analysis of two-dimensional gels. *J. Biol. Chem.* **264**, 5283–5299.

Genbauffe, F.S., Chisholm, G.E., and Cooper, T.G. (1984) Tau, sigma and delta: a family of repeated elements in yeast. *J. Biol. Chem.* **259**, 10518–10525.

Georges, M., Lequarr, A.-S., Castelli, M., Hanset, R., and Vassart, G. (1988) DNA fingerprinting in domestic animals using four different minisatellite probes. *Cyt. Cell. Genet.* **47**, 127–131.

Georges, M., Gunawardana, A., Threadgill, D.W., Lathrop, M. Olsaker, I., Mishra, A., Sargeant, L.L., Schoeberlein, A., Steele, M.R., Terry, C., Threadgill, D.S., Zhao, X., Holm, T., Fries, R., and Womack, J.E. (1991) Characterization of a set of variable number of tandem repeat markers conserved in bovidae. *Genomics* **11**, 24–32.

Gershon, E.S., Martinez, M., Goldin, L.R., and Gejman, P.V. (1990) Genetic mapping of common diseases: the challenges of manic-depressive illness and schizophrenia. *Trends Genet.* **6**, 282–287.

Giannakudis, I., Wolf von Gudenberg, K., Heid, C., and Grzeschik, K.-H. (1992) A series of hypervariable minisatellites at the D7S464 locus. *Hum. Mol. Genet.* **1** (2), 140.

Giannelli, F., Green, P.M., High, K.A., Lozier, J.N., Lillicrap, D.P., Ludwig, M., Olek, K., Reitsma, P.H., Goossens, M., Yoshioka, A., Sommer, S., and Brownlee, G.G. (1990) Haemophilia B: database of point mutations and short additions and deletions. *Nucleic Acids Res.* **18**, 4053–4059.

Gibbs, R.A., and Caskey, C.T. (1987) Identification and localization of mutations at the Lesch–Nyhan locus by ribonuclease A cleavage. *Science* **236**, 303–305.

Gibbs, R.A., Nguyen, P.-N., McBride, L.J., Koepf, S.M., and Caskey, C.T. (1989) Identification of mutations leading to the Lesch–Nyhan syndrome by automated direct DNA sequencing of *in vitro* amplified cDNA. *Proc. Natl. Acad. Sci. USA* **86**, 1919–1923.

Gilbert, W. (1981) DNA sequencing and gene structure (Nobel lecture). *Science* **214**, 1305–1312.

Gilroy, T.E., and Thomas, C.A. (1983) The analysis of some new *Drosophila* repetitive DNA sequences isolated and cloned from two-dimensional agarose gels. Gene 23, 41–51.

Giovannoni, J.J., Wing, R.A., Ganal, M.W., and Tanksley, S.D. (1991) Isolation of molecular markers from specific chromosomal intervals using DNA pools from existing mapping populations. *Nucleic Acids Res.* **19**, 6553–6558.

Gray, M. (1992) Detection of DNA sequence polymorphisms in human genomic DNA by using denaturing gradient gel blots. *Am. J. Hum. Genet.* **50**, 331–346.

Gray, M., Charpentier, A., Walsh, K., Wu, P., and Bender, W. (1991) Mapping point mutations in the Drosophila *rosy* locus using denaturing gradient gel blots. *Genetics* **127**, 139–149.

Grist, S.A., McCarron, M., Kutlaca, A., Turner, D.R., and Morley, A.A. (1992) *In vivo* human somatic mutation: frequency and spectrum with age. *Mutat. Res.* **266**, 189–196.

Grompe, M., Caskey, C.T., and Fenwick, R.G. (1991) Improved molecular diagnostics for ornithine transcarbamylase deficiency. *Am. J. Hum. Genet.* **48**, 212–222.

Gusella, J.F., Keys, C., Varsanyi-Breiner, A., Kao, F.-T., Jones, C., Puck, T.T., and Housman, D. (1980) Isolation and localization of DNA segments from specific human chromosomes. *Proc. Natl. Acad. Sci. USA* **77**, 2829–2833.

Gusella, J.F., Jones, C., Kao, F.-T., Housman, D., and Puck, T.T. (1982) Genetic fine-structure mapping in human chromosome 11 by use of repetitive DNA sequences. *Proc. Natl. Acad. Sci. USA* **79**, 7804–7808.

Haaf, T., and Willard, H.F. (1992) Organization, polymorphism and molecular cytogenetics of chromosome-specific α-satellite DNA from the centromere of chromosome 2. *Genomics* **13**, 122–128.

Haldane, J.B.S. (1935) The rate of spontaneous mutation of a human gene. *J. Genet.* **31**, 317–326.

Hall, J.M., Lee, M.K., Morrow, J., Newman, B., Anderson, L., Huey, B., and King, M.C. (1990) Linkage analysis of early onset familial breast cancer to chromosome 17q21. *Science* **250**, 1684–1689.

Hata, A., Robertson, M., Emi, M., and Lalouel, J.-M. (1990) Direct detection and automated sequencing of individual alleles after electrophoretic strand separation: identification of a common nonsense mutation in exon 9 of the human lipoprotein lipase gene. *Nucleic Acids Res.* **18**, 5407–5411.

Hatada, I., Hayashizaki, Y., Hirotsune, S., Komatsubara, H., and Mukai, T. (1991) A genomic scanning method for higher organisms using restriction sites as landmarks. *Proc. Natl. Acad. Sci. USA* **88**, 9523–9527.

Hearne, C. M., Ghosh, S., and Todd, J.A. (1992) Microsatellites for linkage analyis of genetic traits. *Trends in Genetics* **8**, 288–304.

Herman, G.E., Berry, M., Munro, E., Craig, I.W., and Levy, E.R. (1991) The construction of human somatic cell hybrids containing portions of the mouse X chromosome and their use to generate DNA probes via interspersed repetitive sequence polymerase chain reaction. *Genomics* **10**, 961–970.

Higgs, D.R., Goodbourn, S.E.Y., Wainscoat, J.S., Clegg, J.B., and Weatherall, D.J. (1981) Highly variable regions of DNA flank the human α globin genes. *Nucleic Acids Res.* **9**, 4213–4214.

Higuchi, M., Wong, C., Kochhan, L., Olek, K., Aronis, S., Kasper, C.K., Kazazian, H., and Antonarakis, S.E. (1990) Characterization of mutations in the factor VIII gene by direct sequencing of amplified genomic DNA. *Genomics* **6**, 65–71.

Higuchi, M., Antonorakis, S., Kasch, L., Oldenburg, J., Economou-Petersen, E., Olek, K., Arai, M., Inaba, H., and Kazazian, H.H. (1991) Molecular characterization of mild-to-moderate hemophilia A: detection of the mutation in 25 of 29 patients by denaturing gradient gel electrophoresis. *Proc. Natl. Acad. Sci. USA* **88**, 8307–8311.

Hilbert, P., Lindpaintner, K., Beckmann, J.S., Serikawa, T., Soubrier, F., Dubay, C., Cartwright, P., De Gouyon, B., Julier, C., Takahashi, S., Vincent, M., Ganten, D., Georges, M., and Lathrop, G.M. (1991) Chromosomal mapping of two genetic loci associated with blood-pressure regulation in hereditary hypertensive rats. *Nature* **353,** 521–529.

Hobbs, H.H., Lehrman, M.A., Yamamoto, T., and Russell, D.W. (1985) Polymorphism and evolution of Alu sequences in the human low density lipoprotein receptor gene. *Proc. Natl. Acad. Sci USA* **82,** 7651–7655.

Hofker, M.H., Skraastad, M.I., Bergen, A.A.B., Wapenaar, M.C., Bakker, E., Millington-Ward, A., van Ommen, G.J.B., and Pearson, P.L. (1986) The X-chromosome shows less genetic variation at restriction sites than the autosomes. *Am. J. Hum. Genet.* **39,** 438–451.

Hofmann, S.L., Topham, M., Hsieh, C.-L., and Francke, U. (1991) cDNA and genomic cloning of HRC, a human sarcoplasmic reticulum protein, and localization of the gene to human chromosome 19 and mouse chromosome 7. *Genomics* **9,** 656–669.

Honma, M., Kataoka, E., Ohnishi, K., Kikuno, A., Hayashi, M., Sofuni, T. and Mizusawa, H. (1993) Detection of recombinational mutations in cultured human cells by Southern blot analysis with minisatellite DNA probes. *Mutat. Res.* **286,** 165–172.

Hovig, E., Smith-Sorensen, B., Brogger, A., and Børresen, A.-L. (1991) Constant denaturant gel electrophoresis, a modification of denaturing gradient gel electrophoresis, in mutation detection. *Mutat. Res.* **262,** 63–71.

Hovig, E., Mullaart, E., Børresen, A.-L., Uitterlinden, A. G., and Vijg, J. (1993) Genome scanning of human breast carcinomas using micro- and minisatellite core probes. *Genomics,* **17,** 66–75.

The Huntington's Disease Collaborative Research Group (1993) A novel gene containing a trinucleotide repeat that is expanded and unstable on Huntington's disease chromosomes. *Cell* **72,** 971–983.

Hüvos, P.E., Aquiles Sanchez, J., Kramer, K.M., Karrer, K.M., and Wangh, L.J. (1988) Two-dimensional DNA gel electrophoresis as a method for analysis of eukaryotic genome structure: evaluation using *Tetrahymena thermophilae* DNA. *Biochim. Biophys. Acta* **949,** 325–333.

Iannuzi, M.C., Stern, R.C., Collins, F.S., Tom Hon, C., Hidaka, N., Strong, T., Decker, L., Drumm, M.L., White, M.B., Gerrard, B., and Dean, M. (1991) Two frame-shift mutations in the cystic fibrosis gene. *Am. J. Hum. Genet.* **48,** 227–231.

Ionov, Y., Peinado, M.A., Malkhosyan, S., Shibata, D., and Perucho, M. (1993) Ubiquitous somatic mutations in simple repeated sequences reveal a new mechanism for colonic carcinogenesis. *Nature* **363,** 558–561.

Isobe, M., Emanuel, B.S., Givol, D., Oren, M., and Croce, C.M. (1986) Localisation of the gene for human p53 tumour antigen to band 17p13. *Nature* **320,** 84–85.

Ivaschenko, T.E., White, M.B., Dean, M., and Baranov, V.S. (1991) A deletion of two nucleotides in exon 10 of the CFTR gene in a Soviet family with cystic fibrosis causing early infant death. *Genomics* **10,** 298–299.

Jarman, A.P., Nicholls, R.D., Weatherall, D.J., Clegg, J.B., and Higgs, D.R. (1986) Molecular characterisation of a hypervariable region downstream of the human alpha-globin gene cluster. *EMBO J.* **5**, 1857–1863.

Jeffreys, A.J. (1979) DNA sequence variants in the G_γ, A_γ,- and β-globin genes of man. *Cell* **18**, 1–15.

Jeffreys, A.J. (1985) Method of characterizing a test sample of genomic DNA. *European Patent 186271.*

Jeffreys, A.J., Wilson, V., and Thein, S.L. (1985a) Hypervariable 'minisatellite' regions in human DNA. *Nature* **314**, 67–73.

Jeffreys, A.J., Wilson, V., and Thein, S.L. (1985b) Individual-specific 'fingerprints' of human DNA. *Nature* **316**, 76–79.

Jeffreys, A.J., Wilson, V., Thein, S.L., Weatherall, D.J., and Ponder, B.A.J. (1986) DNA 'fingerprints' and segregation analysis of multiple markers in human pedigrees. *Am. J. Hum. Genet.* **39**, 11–24.

Jeffreys, A.J., Royle, N.J., Wilson, V., and Wong, Z. (1988) Spontaneous mutation rates to new length alleles at tandem-repetitive hypervariable loci in human DNA. *Nature* **332**, 278–281.

Jeffreys, A.J., Neumann, R., and Wilson, V. (1990) Repeat unit sequence variation in minisatellites: novel source of DNA polymorphism for studying variation and mutation by single molecule analysis. *Cell* **60**, 473–485.

Jeffreys, A.J., Turner, M., and Debenham, P. (1991a) The efficiency of multilocus DNA fingerprint probes for individualization and establishment of family relationships, determined from extensive casework. *Am. J. Hum. Genet.* **48**, 824–840.

Jeffreys, A.J., MacLeod, A., Tamaki, K., Neil, D.L., and Monckton, D.G. (1991b) Minisatellite repeat coding as a digital approach to DNA typing. *Nature* **354**, 204–209.

Julier, C., de Gouyon, B., Georges, M., Guénet, J.-L., Nakamura, Y., Avner, P., and Lathrop, G.M. (1990) Minisatellite linkage maps in the mouse by cross-hybridization with human probes containing tandem repeats. *Proc. Natl. Acad. Sci. USA* **87**, 4585–4589.

Kaback, D.B., Steensma, Y.H., and Jonge, P. de (1989) Yeast shows higher ratio of genetic to physical distance. *Proc. Natl. Acad. Sci. USA* **86**, 3694–3698.

Kallioniemi, A., Kallioniemi, O.-P., Sudar, D., Rutovitz, D., Gray, J.W., Waldman, F., and Pinkel, D. (1992) Comparative genomic hybridization for molecular cytogenetic analysis of solid tumors. *Science* **258**, 818–821.

Kamiura, S., Nolan, C.M., and Meruelo, D. (1992) Long-range physical map of the Ly-6 complex: mapping the Ly-6 multigene family by field-inversion and two-dimensional gel electrophoresis. *Genomics* **12**, 89–105.

Kan, Y.W. and Dozy, A.M. (1978) Polymorphisms of DNA sequence adjacent to human β-globin structural gene: Relationship to sickle mutation. *Proc. Natl. Acad. Sci. USA* **75**, 5631–5635.

Kashi, Y., Tikochinsky, Y., Iraqi, F., Nave, A., Beckman, J.S., Gruenbaum, Y., and Soller, M. (1990) Large restriction fragments containing poly-TG are highly polymorphic in a variety of vertebrates. *Nucleic Acids Res.* **18**, 1129–1132.

Katzir, N., Rechavi, G., Cohen, J.B., Unger, T., Simoni, F., Segal, S., Cohen, D., and Givol, D., (1985) 'Retroposon' insertion into the cellular oncogene c-*myc* in canine transmissible venereal tumor. *Proc. Natl. Acad. Sci. USA* **82,** 1054–1058.

Kazazian, H.H., Wong, C., Youssoufian, H., Scott, A.F., Phillips, D.G., and Antono-rakis, S.E. (1988) Haemophilia A resulting from *de novo* insertion of L1 sequences represents a novel mechanism for mutation in man. *Nature* **332,** 164–166.

Keen, T.J., Inglehearn, C.F., Lester, D.H., Bashir, R., Jay, M., Bird, A.C., Jay, B., and Bhattacharya, S.S. (1991) Autosomal dominant retinis pigmentosa: four new mutations in rhodopsin, one of them in the retinal attachment site. *Genomics* **11,** 199–205.

Kenwrick, S., Patterson, M., Speer, A., Fischbeck, K., and Davies, K.E. (1987) Molecular analysis of the Duchenne muscular dystrophy region using pulsed field gel electrophoresis. *Cell* **48,** 351–357.

Keohavong, P., Liu, V.F., and Thilly, W.G. (1991) Analysis of point mutations induced by ultraviolet light in human cells. *Mutat. Res.* **249,** 147–159.

Kern, S.E., Fearon, E.R., Tersmette, K.W.F., Enterline, J.P., Leppert, M., Nakamura, Y., White, R., Vogelstein, B., and Hamilton, S.R. (1989) Allelic loss in colorectal carcinoma. *J. Am. Med. Assoc.* **261,** 3099–3103.

Kikuno, T., Honma, M., Kataoka, E., Ohnishi, K., Inoue, A., Mizusawa, H., and Sofuni, T. (1992) Detection of minisatellite DNA mutations induced by MNNG treatment with DNA fingerprint analysis. *Mutat. Res.* **272,** 267.

Kipling, D., and Cooke, H.J. (1992) Beginning or end? Telomere structure, genetics and biology. *Hum. Mol. Genet.* **1,** 3–6.

Knott, T.J., Wallis, S.C., Pease, R.J., Powell, L.M., and Scott, J. (1986) A hypervariable region 3' to the human apolipoprotein B gene. *Nucleic Acids Res.* **14,** 9215–9216.

Knowlton, R.G., Nelson, C.A., Brown, V.A., Page, D.C., and Donis-Keller, H. (1989) An extremely polymorphic locus on the short arm of the human X chromosome with homology to the long arm of the Y chromosome. *Nucleic Acids Res.* **17** (1), 423–437.

Koeberl, D.D., Bottema, C.D., Ketterling, R.P., Bridge, P.J., Lillicrap, D.P., and Sommer, S.S. (1990) Mutations causing hemophilia B: direct estimate of the underlying rates of spontaneous germ-line transitions, transversions, and deletions in a human gene. *Am. J. Hum. Genet.* **47,** 202–217.

Koenig, M., Hoffman, E.P., Bertelson, C.J., Monaco, A.P., Feener, C., and Kunkel, L.M. (1987) Complete cloning of the Duchenne muscular dystrophy (DMD) cDNA and preliminary genomic organization of the DMD gene in normal and affected individuals. *Cell* **50,** 509–517.

Kok, K., Osinga, J., Carritt, B., Davis, M.B., van der Hout, A.H., van der Veen, A.Y., Landsvater, R.M., de Leij, L.F.M.H., Berendsen, H.H., Postmus, P.E., Poppema, S., and Buys, C.H.C.M. (1987) Deletion of a DNA sequence at the chromosomal region 3p21 in all major types of lung cancer. *Nature* **330,** 578–581.

Kominami, R., Urano, Y., Mishima, Y., and Muramatsu, M. (1983a) Novel repetitive sequence families showing size and frequency polymorphisms in the genomes of mice. *J. Mol. Biol.* **165,** 209–228.

Kominami, R., Muramatsu, M., and Moriwaki, K. (1983b) A mouse type 2 Alu sequence (M2) is mobile in the genome. *Nature* **301,** 87–89.

Kovar, H., Jug, G., Auer, H., Skern, T., and Blaas, D. (1991) Two-dimensional single-strand conformation polymorphism analysis: a useful tool for the detection of mutations in long DNA fragments. *Nucleic Acids Res.* **19** (13), 3507–3510.

Krolewski, A.S., Keolewski, B., Gray, M., Stanton, V., Warram, J.H., and Housman, D. (1992) High-frequency DNA sequence polymorphisms in the insulin receptor gene detected by denaturing gradient gel blots. *Genomics* **12,** 705–709.

Kuff, E.L., and Leuders, K.K. (1988) Intracisternal A-particle gene family: structural and functional aspects. *Adv. Cancer Res.* **51,** 183–276.

Labrune, P., Melle, D., Rey, F., Berthelon, M., Caillaud, C., Rey, J., Munnich, A., and Lyonnet, S. (1991) Single-strand conformation polymorphism for detection of mutations and base substitutions in phenylketonuria. *Am. J. Hum. Genet.* **48,** 1115–1120.

Lakshmikumaran, M.S., D'Ambrosio, E., Laimins, L.A., Lin, D.T., and Furano, A.V. (1985) Long interspersed repeated DNA (LINE) causes polymorphism at the rat insulin 1 locus. *Mol. Cell. Biol.* **5,** 2197–2203.

Lalouel, J.-M., Lathrop, G.M., and White, R. (1986) Construction of human genetic linkage maps: II. Methodological issues. In: *Molecular Biology of Homo Sapiens.* Cold Spring Harbor Laboratory Press, New York, pp. 39–47.

Lander, E. S. (1988) Mapping complex genetic traits in humans. In: *Genome Analysis, a practical approach* (ed. Davies, K.E.) IRL Press, Oxford, pp. 171–189.

Lander, E.S., and Botstein, D. (1986) Mapping complex genetic traits in humans: new methods using a complete RFLP linkage map. In: *Molecular Biology of Homo sapiens.* Cold Spring Harbor Laboratory Press, New York, pp. 49–62.

Lander, E.S., and Botstein, D. (1989) Mapping Mendelian factors underlying quantitative traits using RFLP linkage maps. *Genetics* **121,** 185–199.

Latham, T., and Smith, F. (1989) Detection of single-base mutations in DNA molecules using the solution melting method. *DNA* **8,** 223–231.

Lathrop, G.M., and Lalouel, J.M. (1984) Easy calculation of lod scores and genetic risks on small computers. *Am. J. Hum. Genet.* **36,** 460–465.

Lawrance, S.K., Smith, C.L., Srivastava, R., Cantor, C.R., and Weissman, S.M. (1987) Megabase-scale mapping of the HLA gene complex by Pulsed Field gel electrophoresis. *Science* **235,** 1387–1390.

Ledbetter, S.A., Nelson, D.L., Warren, S.T., and Ledbetter, D.H. (1990a) Rapid isolation of DNA probes within specific chromosome regions by interspersed repetitive sequence polymerase chain reaction. *Genomics* **6,** 475–481.

Ledbetter, S.A., Garcia-Heras, J., and Ledbetter, D.H. (1990b) 'PCR-karyotype' of human chromosomes in somatic cell hybrids. *Genomics* **8,** 614–622.

Lehrach, H., Drmanac, R., Hoheisel, J., Larin, Z., Lennon, G., Monaco, A.P., Nizetic, D., Zehetner, G., and Poustka, A. (1990) Hybridization fingerprinting in genome mapping and sequencing. In: Davies, K.E., and Tilghman, S. M. eds, *Genome Analysis* Vol. 1, *Genetic and Physical Mapping.* Cold Spring Harbor Laboratory Press, New York, pp. 39–81.

Lehrman, M.A., Goldstein, J.L., Russell, D.W., and Brown, M. (1987) Duplication of seven exons in LDL receptor gene caused by Alu–Alu recombination in a subject with familial hypercholesterolemia. *Cell* **48,** 827–835.

Leider, J.M., Palese, P., and Smith, F.I. (1988) Determination of the mutation rate of a retrovirus. *J. Virol.* **62**, 3084–3091.

Lennon, G.C., and Lehrach, H. (1991) Hybridization analysis of arrayed cDNA libraries. *Trends Genet.* **7** (10), 314–317.

Lerman, L.S., Fischer, S.G., Bregman, D.B., and Silverstein, K.J. (1981) Base sequence and melting thermodynamics determine the position of pBR322 fragments in two-dimensional gel electrophoresis. In: Sarma, R. H. (ed.), *Biomolecular stereodynamics*. Adenine Press, Albany, NY, pp. 459–470.

Lerman, L.S., Fischer, S.G., Hurley, I., Silverstein, K., and Lumelsky, N. (1984) Sequence-determined DNA separations. *Annu. Rev. Biophys. Bioeng.* **13**, 399–423.

Lerman, L.S., Silverstein, K., and Grinfield, E. (1986) Searching for gene defects by denaturing gradient gel electrophoresis. *Cold Spring Harbor Symp. Quant. Biol.* **51**, 285–297.

Levine, A.J. (1992) The p53 tumor-suppressor gene. *N. Engl. J. Med.* **326**, 1350–1352.

Liang, P., and Pardee, A.B. (1992) Differential display of eukaryotic messenger RNA by means of the polymerase chain reaction. *Science* **257**, 967–971.

Lichter, P., Ledbetter, S.A., Ledbetter, D.H., and Ward, D.C. (1990) Fluorescence *in situ* hybridization with Alu I and L1 polymerase chain reaction probes for rapid characterization of human chromosomes in hybrid cell lines. *Proc. Natl. Acad. Sci. USA* **87**, 6634–6638.

Lillie, J.W., Green, M., and Green, M.R. (1986) An adenovirus E1a protein region required for transformation and transcriptional repression. *Cell* **46**, 1043–1051.

Lin, C.S., Goldthwait, D.A., and Samols, D. (1988) Identification of Alu transposition in human lung carcinoma cells. *Cell* **54**, 153–159.

Lisitsyn, N., Lisitsyn, N., and Wigler, M. (1993) Cloning the differences between two complex genomes. *Science* **259**, 946–951.

Litt, M., and Luty, J.A. (1989) A hypervariable microsatellite revealed by *in vitro* amplification of a dinucleotide repeat within the cardiac muscle actin gene. *Am. J. Hum. Genet.* **44**, 397–401.

Losekoot, M., Fodde, R., Harteveld, C.L., van Heeren, H. Giordano P.C., Bernini, L.F. (1990) Denaturing gradient gel electrophoresis and direct sequencing of PCR amplified genomic DNA: a rapid and reliable diagnostic approach to β-thalassemia. *Br. J. Haematol.* **76**, 269–274.

Luca Cavalli-Sforza, L.(1990) Opinion: how can one study individual variation for 3 billion nucleotides of the human genome? *Am. J. Hum. Genet.* **46**, 649–651.

Luty, J.A., Guo, Z., Willard, H.F., Ledbetter, D.H., Ledbetter, S., and Litt, M. (1990) Five polymorphic microsatellite VNTRs on the human X chromosome. *Am. J. Hum. Genet.* **46**, 776–783.

Maciejko, D., Bal, J., Mazurczak, T., te Meerman, G., Buys, C., Oostra, B., and Halley, D. (1989) Different haplotypes for cystic fibrosis-linked DNA polymorphisms in Polish and Dutch populations. *Hum. Genet.* **83**, 220–222.

Mackay, J., Steel, C.M., Elder, P.A., Forrest, A.P.M., and Evans, H.J. (1988) Allele loss on short arm of chromosome 17 in breast cancers. *Lancet* **ii**, 1384–1385.

Maddalena, A., Spence, J.E., O'Brien, W.E., and Nussbaum, R.L. (1988) Characterization of point mutations in the same arginine codon in three unrelated patients with ornithine transcarbamoylase deficiency. *J. Clin. Invest.* **82,** 1353–1358.

Mager, D.L., and Goodchild, N.L. (1989) Homologous recombination between the LTRs of a human retrovirus-like element causes a 5-kb deletion in two siblings. *Am. J. Hum. Genet.* **45,** 848–854.

Mahadevan, M., Tsilfidis, C., Sabourin, L., Shutler, G., Amemiya, C., Jansen, G., Neville, C., Narang, M., Barcelo, J., O'Hoy, K., Leblond, S., Earle-MacDonald, J., de Jong, P.J., Wieringa, B., Korneluk, R.G. (1992) Myotonic dystrophy mutation: an unstable CTG repeat in the 3' untranslated region of the gene. *Science* **255,** 1253–1255.

Maier, E., Hoheisel, J.D., McCarthy, L., Mott, R., Grigoriev, A.V., Monaco, A.P., Larin, Z., and Lehrach, H. (1992) Complete coverage of the *Schizosaccharomyces pombe* genome in yeast artificial chromosomes. *Nature Genet.* **1,** 273–277.

Mariat, D., and Vergnaud, G. (1992) Detection of polymorphic loci in complex genomes with synthetic tandem repeats. *Genomics* **12,** 454–458.

Marx, J.M. (1990) Dissecting the complex diseases. *Science* **247,** 1540–1544.

Mathies, R.A. and Huang, X.C. (1992) Capillary array electrophoresis: an approach to high-speed, high-throughput DNA sequencing. *Nature* **359,** 167–169.

Maxam, A.M., and Gilbert, W. (1977) A new method of sequencing DNA. *Proc. Natl. Acad. Sci. USA* **74,** 560–564.

McClintock, B. (1984) The significance of responses of the genome to challenge. *Science* **226,** 792–801.

McGuire, W.L. and Naylor, S.L. (1989) Loss of heterozygosity in breast cancer: cause or effect? *J. Natl. Cancer Inst.* **81,** 1764–1765.

McKusick, V.A. (1990) *Mendelian Inheritance in Man.* 9th edn. Johns Hopkins University Press, Baltimore, MD.

te Meerman, G.J. (1991) A logic programming approach to pedigree analysis. PhD thesis, University of Groningen (ISBN 90 5170 0776).

te Meerman, G.J., Mullaart, E., van der Meulen, M. A., den Daas, J.H.G., Morolli, B., Uitterlinden, A.G., and Vijg, J. (1993) Linkage analysis by two-dimensional DNA typing. *Am. J. Hum. Genet.* **53,** 1289–1297.

Meese, E.U., Meltzer, P.S., Ferguson, P.W., and Trent, J.M. (1992) Alu-PCR: characterization of a chromosome 6-specific hybrid mapping pannel and cloning of chromosome-specific markers. *Genomics* **12,** 549–554.

Melchior, W.B., and Von Hippel, P.H. (1973) Alteration of the relative stability of dA.dT and dG.dC base pairs in DNA. *Proc. Natl. Acad. Sci. USA* **70,** 298–302.

Melmer, G. and Buchwald, M. (1992) Identification of genes using oligonucleotides corresponding to splice site consensus sequences. *Hum. Mol. Genet.* **1,** 433–438.

Meulenbelt, I.M., Wapenaar, M.C., Patterson, D., Vijg, J., and Uitterlinden, A.G. (1993) Isolation and mapping of human chromosome 21 cosmids using a probe for RTVL-H retrovirus-like elements. *Genomics,* **15,** 492–499.

Meyer, G., and Hildebrandt, A. (1986) Two-dimensional gel analysis of repetitive nuclear DNA sequences in the genome of *Physarum polycephalum. Eur. J. Biochem.* **157**, 507–512.

Michelmore, R.W., Paran, I., and Kesseli, R.V. (1991) Identification of markers linked to disease-resistance genes by bulked segregant analysis: a rapid method to detect markers in specific genomic regions by using segregating populations. *Proc. Natl. Acad. Sci. USA* **88**, 9828–9832.

Mietz, J.A., and Kuff, E.L. (1992) Intracisternal A-particle specific oligonucleotides provide multilocus probes for genetic linkage studies in the mouse. *Mamm. Genome* **3**, 447–451.

Mitelman, F. (1991) *Catalog of Chromosome Aberrations in Cancer.* 4th edn, Wiley–Liss, New York.

Monaco, A.P., Neve, R., Colletti-Feener, C., Bertelson, C.J., Kurnit, D.M., and Kunkel, L.M. (1986) Isolation of candidate cDNAs for portions of the Duchenne muscular dystrophy gene. *Nature* **323**, 646–648.

Montandon, A.J., Green, P.M., Giannelli, F., and Bently, D.R. (1989) Direct detection of point mutations by mismatch analysis: application to haemophilia B. *Nucleic Acids Res.* **17**, 3347–3358.

Morell, V. (1993) Huntington's gene finally found. *Science* **260**, 28–30.

Moroz-Williamson, V. (1983) Transposable elements in yeast. *Int. Rev. Cytol.* **83**, 1–24.

Mullaart, E., de Vos, G.J., te Meerman, G.J., Uitterlinden, A.G., and Vijg, J. (1993) Parallel genome analysis by two-dimensional DNA typing. *Nature* **365**, 469–471.

Mullis, K.B., and Faloona, F. (1987) Specific synthesis of DNA *in vitro* via a polymerase catalysed chain reaction. In: Wu, R. (ed.) *Methods in Enzymology.* Vol. 155. Academic Press, New York, pp. 335–350.

Musich, P.R., and Dykes, R.J. (1986) A long interspersed (LINE) DNA exhibiting polymorphic patterns in human genomes. *Proc. Natl. Acad. Sci. USA* **83**, 4854–4858.

Muyzer, G., de Waal, E.C., and Uitterlinden, A.G. (1993) Profiling of complex microbial populations by denaturing gradient gel electrophoresis of polymerase chain reaction amplified genes coding for 16S ribosomal DNA. *Appl. Environ. Microbiol.,* **59**, 695–700.

Myers, R.M., Lumelsky, N., Lerman, L.S., and Maniatis, T. (1985a) Detection of single base substitutions in total genomic DNA. *Nature* **313**, 495–498.

Myers, R.M., Fischer, S.G., Maniatis, T., and Lerman, L.S. (1985b) Modification of the melting properties of duplex DNA by attachment of a GC-rich DNA sequence as determined by denaturing gradient gel electrophoresis. *Nucleic Acids Res.* **13**, 3111–3129.

Myers, R.M., Fischer, S.G., Lerman, L.S., and Maniatis, T. (1985c) Nearly all single base substitutions in DNA fragments joined to a GC-clamp can be detected by denaturing gradient gel electrophoresis. *Nucleic Acids Res.* **13**, 3131–3145.

Myers, R.M., Lerman, L.S., and Maniatis, T. (1985d) A general method for saturation mutagenesis of cloned DNA fragments. *Science* **229**, 242–247.

Myers, R.M., Larin, Z., and Maniatis, T. (1985e) Detection of single base substitutions by ribonuclease cleavage at mismatches in RNA:DNA duplexes. *Science* **230**, 1242–1246.

Myers, R.M., Sheffield, V.C., and Cox, D.R. (1989) Detection of single base changes in DNA: ribonuclease cleavage and denaturing gradient gel electrophoresis. In: Davies, K. E. (ed.), *Genome analysis: A Practical Approach* IRL Press, Oxford, 95–139.

Nakamura, Y., Leppert, M., O'Connell, P., Wolff, R., Holm, T., Culver, M., Martin, C., Fujimoto, E., Hoff, M., Kumlin, E., and White, R. (1987a) Variable Number of Tandem Repeat (VNTR) markers for human gene mapping. *Science* **235**, 1616–1622.

Nakamura, Y., Julier, C., Wolff, R., Holm, T., O'Connell, P., Leppert, M., and White, R. (1987b) Characterization of a human *midisatellite* sequence. *Nucleic Acids Res.* **15**, 2537–2547.

Nakamura, Y., Carlson, M., Krapcho, K., Kanamori, M., and White, R. (1988a) New approach for isolation of VNTR markers. *Am. J. Hum. Genet.* **43**, 854–859.

Nakamura, Y., Lathrop, M., O'Connell, P., Leppert, M., Barker, D., Wright, E., Skolnick, M., Kondoleon, S., Litt, M., Lalouel, M., and White, R. (1988b) A mapped set of DNA markers for human chromosome 17. *Genomics* **2**, 302–309.

Neel, J.V., Satoh, C., and Myers, R. (1993) Report of a workshop on the application of molecular genetics to the study of mutation in the children of atomic bomb survivors. *Mutat. Res.* **291**, 1–20.

Nelson, D.L., Ledbetter, S., Corbo, L., Victoria, M.F., Ramirez-Solis, R, Webster, T.D., Ledbetter, D.H., and Caskey, C.T. (1989) Alu polymerase chain reaction: A method for rapid isolation of human-specific sequences from complex DNA sources. *Proc. Natl. Acad. Sci. USA* **86**, 6686–6690.

Nelson, S.F., McCusker, J.H., Sander, M.A., Kee, Y., Mpdrich, P., and O'Brown, P.O. (1993) Genomic mismatch scanning: a new approach to genetic linkage mapping. *Nature Genet.* **4**, 11–18.

Newton, C.R., Graham, A., Heptinstall, L.E., Powell, S.J., Summers, C., Kalsheker, N., Smith, J.C., and Markham, A.F. (1989) Analysis of any point mutation in DNA. The amplification refractory mutation system (ARMS). *Nucleic Acids Res.* **17**, 2503–2516.

Nigro, J.M., Baker, S.J., Preisinger, A.C., Jessup, J.M., Hostetter, R., Cleary, K., Bigner, S., Davidson, N., Baylin, S., Devilee, P., Glover, T., Collins, F.S., Weston, A., Modali, R., Harris, C.C., and Vogelstein, B. (1989) Mutations in the p53 gene occur in diverse human tumor types. *Nature* **342**, 705–708.

Nizetic, D., Zehetner, G., Monaco, A. P., Gellen, L., Young, B. D., and Lehrach, H. (1991) Construction, arraying, and high-density screening of large insert libraries of human chromosomes X and 21: their potential use as reference libraries. *Proc. Natl. Acad. Sci. USA* **88**, 3233–3237.

NIH/CEPH Collaborative Mapping Group (1992) A comprehensive genetic linkage map of the human genome. *Science* **258**, 67–86.

Noll, W.W., and Collins, M. (1987) Detection of human DNA polymorphisms with a simplified denaturing gradient gel electrophoresis technique. *Proc. Natl. Acad. Sci. USA* **84**, 3339–3343.

Novack, D.F., Casna, N.J., Fischer, S.G., and Ford, J. (1986) Detection of single base-pair mismatches in DNA by chemical modification followed by electrophoresis in 15% polyacrylamide gel. *Proc. Natl. Acad. Sci. USA* **83**, 586–590.

Nürnberg, P., Roewer, L., Neitzel, H., Sperling, K., Pöpperl, A., Hundrieser, J., Pöche, H., Epplen, C., Zischler, H., and Epplen, J.T. (1989) DNA fingerprinting with the oligonucleotide probe $(CAC)_5/(GTG)_5$: somatic stability and germline mutations. *Hum. Genet.* **84**, 75–78.

Oliver, S.G., van der Aart, Q.J.M., Agostoni-Carbone, M.L., Aigle, M., and 143 others (1992) The complete DNA sequence of yeast chromosome III. *Nature* **357**, 38–47.

Olson, M.V., Hood, L., Cantor, C.R., and Botstein, D. (1989) A common language for physical mapping of the human genome. *Science* **245**, 1434–1435.

Orita, M., Iwahana, H., Kanazawa, H., Hayashi, K., and Sekiya, T. (1989a) Detection of polymorphisms of human DNA by gel electrophoresis as single-strand conformation polymorphisms. *Proc. Natl. Acad. Sci. USA* **86**, 2766–2770.

Orita, M., Suzuki, Y., Sekiya, T., and Hayashi, K. (1989b) Rapid and sensitive detection of point mutations and DNA polymorphisms using the polymerase chain reaction. *Genomics* **5**, 874–879.

Orkin, S.H. (1986) Reverse Genetics and human disease. *Cell* **47**, 845–846.

Orkin, S.H., Little, P.F.R., Kazazian, H.H.,Jr, and Boehm, C.D. (1982) Improved detection of the sickle mutation by DNA analysis: application to prenatal diagnosis. *N.Engl. J. Med.* **307**, 32–36.

Ott, J. (1991) *Analysis of Human Genetic Linkage,* 2nd edn, Johns Hopkins University Press, Baltimore, MD.

Page, D.C., Bieker, K., Brown, L.G., Hinton, S., Leppert, M., Lalouel, J.-M., Lathrop, M., Nystrom-Lahti, M., de la Chapelle, A., and White, R. (1987) Linkage, physical mapping, and DNA sequence analysis of pseudoautosomal loci on the human X and Y chromosome. *Genomics* **1**, 243–256.

Paterson, A.H., Lander, E.S., Hewitt, J.D., Peterson, S., Lincoln, S.E., and Tanksly, S.D. (1988) Resolution of quantitative traits into Mendelian factors by using a complete linkage map of restriction fragment length polymorphisms. *Nature* **335**, 721–726.

Paw, B.H., Tieu, P.T., Kaback, M.M., Lim, J., and Neufeld, E.F. (1990) Frequency of three Hex A mutant alleles among Jewish and non-Jewish carriers identified in a Tay–Sachs screening program. *Am. J. Hum. Genet.* **47**, 698–705.

Peltomäki, P., Aaltonen, L.A., Sistonen, P., Pylkänen, L., Mecklin, J.-L., Järvinen, H., Green, J.S., Jass, J.R., Weber, J.L., Leach, F.S., Petersen, G.M., Hamilton, S.R., de la Chapelle, A., and Vogelstein, B. (1993) Genetic mapping of a locus predisposing to human colorectal cancer. *Science* **260**, 810–812.

Piratsu, M., Kan, Y.W., Cao, A., Conner, B.J., Teplitz, R.L., and Wallace, R.B. (1983) Prenatal diagnosis of β-thalassemia: detection of a single nucleotide mutation in DNA. *N. Engl. J. Med.* **309**, 284–287.

Po, T., Steger, G., Rosenbaum, V., Kaper, J., and Riesner, D. (1987) Double-stranded cucumovirus associated RNA 5: experimental analysis of necrogenic and non-necrogenic variants by temperature-gradient gel electrophoresis. *Nucleic Acids Res.* **15**, 5069–5083.

Poddar, S.K., and Maniloff, J. (1986) Chromosome analysis by two-dimensional fingerprinting. *Gene* **49**, 93–102.

Poddar, S.K., and Maniloff, J. (1989) Determination of microbial genome sizes by two-dimensional denaturing gradient gel electrophoresis. *Nucleic Acids Res.* **17**, 2889–2895.

Poduslo, S.E., Dean, M., Kolch, U., and O'Brien, S.J. (1991) Detecting high-resolution polymorphisms in human coding loci by combining PCR and single-strand conformation polymorphism (SSCP) analysis. *Am. J. Hum. Genet.* **49**, 106–111.

Ponder, B. (1988) Gene losses in human tumours. *Nature* **335**, 400–403.

Porteous, D.J., Morten, J.E.N., Cranston, G., Fletcher, J.M., Mitchell, A., van Heyningen, V., Fantes, J.A., Boyd, P.A., and Hastie, N.D. (1986) Molecular and physical arrangements of human DNA in HRAS1-selected, chromosome-mediated transfectants. *Mol. Cell. Biol.* **6**, 2223–2232.

Potter, S.S., Bott, K.F., and Newbold, J.E. (1977) Two dimensional restriction analysis of the *Bacillus subtilis* genome: gene purification and ribosomal ribonucleic acid gene organization. J. Bacteriol. 129, 492–500.

Poustka, A., and Lehrach, H. (1986) Jumping libraries and linking libraries: the next generation of molecular tools in mammalian genetics. *Trends Genet.* **2**, 174–179.

Premkumar Reddy, E., Reynolds, R.K., Santos, E., and Barbacid, M. (1982) A point mutation is responsible for the acquisition of transforming properties by the T24 human bladder carcinoma oncogene. *Nature* **300**, 149–152.

Proudfoot, N.J., Gil, A., and Maniatis, T. (1982) The structure of the human zeta-globin gene and a closely linked, nearly identical pseudogene. *Cell* **31**, 553–563.

Riedel, G.E., Swanberg, S.L., Kurunda, K.D., Marquette, K., LaPan, P., Bledsoe, P., Kennedy, A., and Lin, B.-Y. (1990) Denaturing gradient gel electrophoresis identifies genomic DNA polymorphism with high frequency in maize. *Theor. Appl. Genet.* **80**, 1–10.

Riedy, M.F., Hamilton III, W.J., and Aquadro, C.F. (1992) Excess of non-parental bands in offspring from known primate pedigrees assayed using RAPD PCR. *Nucleic Acids Res.* **20**, 918.

Riggins, G.J., Lokey, L.K., Chastain, J.L., Leiner, H.A., Sherman, S.L., Wilkinson, K.D., and Warren, S.T. (1992) Human genes containing polymorphic trinucleotide repeats. *Nature Genet.* **2**, 186–191,

Riordan, J.R., Rommens, J.M., Kerem, B.-S., Alon, N., Rozmahel, R., Grzelczak, Z., Zielenski, J., Lok, S., Plavsic, N., Chou, J.-L., Drumm, M.L., Ianuzzi, M.C., Collins, F.S., and Tsui, L.-C. (1989) Identification of the cystic fibrosis gene: cloning and characterization of complementary DNA. *Science* **245**, 1066–1073.

Roberts, L. (1990a) Huntington's gene: so near, yet so far. *Science* **247**, 624–627.

Roberts, L. (1990b) An animal genome project? *Science* **248**, 550–552.

Roberts, R.G., Bobrow, M., and Bently, D.R. (1992) Point mutations in the dystrophin gene. *Proc. Natl. Acad. Sci. USA* **89**, 2331–2335.

Rogan, P.K., Lemkin, P.L., Klar, A.J.S., Singh, J., and Strathern, J.N. (1991) Two-dimensional agarose gel electrophoresis of restriction-digested genomic DNA. *Methods: Companion Methods Enzymol.* **3** (2), 91–97.

Rogstad, S.H., Patton II, J.C., and Schaal, B.A. (1988) M13 repeat probe detects DNA minisatellite-like sequences in gymnosperms and angiosperms. *Proc. Natl. Acad. Sci. USA* **85**, 9176–9178.

Rogstad, S.H., Herwaldt, B.L., Schlesinger, P.H., and Krogstad, D.J. (1989) The M13 repeat probe detects RFLPs between two strains of the protozoan malaria parasite *Plasmodium falciparum. Nucleic Acids Res.* **17**, 3610.

Rommens, J.M., Iannuzzi, M.C., Kerem, B.-S., Drumm. M.L., Melmer, G., Dean, M., Rozmahel, R., Cole, J.L., Kennedy, D., Hidaka, N., Zsiga, M., Buchwald, M., Riordan, J.R., Tsui, L.-P., and Collins, F.S. (1989) Identification of the cystic fibrosis gene: chromosome walking and jumping. *Science* **245**, 1059–1065.

Rosatelli, M.C., Dozy, A., Faà, V., Meloni, A., Sardu, R., Saba, L., Kan, Y.W., and Cao, A. (1992) Molecular characterization of β-thalassemia in the Sardinian population. *Am. J. Hum. Genet.* **50**, 422–426.

Rosenbaum, V., and Riesner, D. (1987) Temperature gradient electrophoresis. Thermodynamic analysis of nucleic acids and proteins in purified form and in cellular extracts. *Biophys. Chem.* **26**, 235–246.

Royle, N.J., Clarkson, R.E., Wong, Z., and Jeffreys, A.J. (1988) Clustering of hypervariable minisatellites in the proterminal regions of human autosomes. *Genomics* **3**, 352–360.

Runnebaum, I.B., Nagarajan, M., Bowman, M., Soto, D., and Sukumar, S. (1991) Mutations in p53 as potential molecular markers for human breast cancer. *Proc. Natl. Acad. Sci. USA* **88**, 10657–10661.

Ryskov, A.P., Jincharadze, A.G., Prosnyak, M.I., Ivanov, P.L., and Limborska, S.A. (1988) M13 phage as a universal marker for DNA fingerprinting of animals, plants and microorganisms. *FEBS Lett.* **233**, 388–392.

Sager, R. (1989) Tumor suppressor genes: the puzzle and the promise. *Science* **246**, 1406–1412.

Saiki, R.K., Scharf, S., Faloona, F., Mullis, K.B., Horn, G., Erlich, H.A., and Arnheim, N. (1985) Enzymatic amplification of β-globin genomic sequences and restriction site analysis for diagnosis of sickle cell anemia. *Science* **230**, 1350–1354.

Saiki, R.K., Bugawan, T.L., Horn, G.T., Mullis, K.B., and Erlich, H.A. (1986) Analysis of enzymatically amplified β-globin and HLA-DQ DNA with allele-specific oligonucleotide probes. *Nature* **324**, 163–166.

Saiki, R.K., Chang, C.A., Levenson, C.H., Warren, T.C., Boehm, C.D., Kazazian, H.H., Jr, and Erlich, H.A. (1988) Diagnosis of sickle cell anemia and β-thalassemia with enzymatically amplified DNA and non-radioactive allel-specific oligonucleotide probes. *N. Engl. J. Med.* **319**, 537–541.

Saiki, R.K., Walsh, P.S., Levenson, C.H., and Erlich, H.A. (1989) Genetic analysis of amplified DNA with immobilized sequence-specific oligonucleotide probes. *Proc. Natl. Acad. Sci. USA* **86**, 6230–6234.

Sakaki, Y., Kurata, Y., Miyake, T., and Saigo, K. (1983) Two dimensional gel electrophoretic analysis of the Hind III 1.8 kb repetitive-sequence family in the human genome. *Gene* **24**, 179–190.

Sambrook, J., Fritsch, E.F., and Maniatis, T. (1989) *Molecular Cloning: A Laboratory Manual.* 2nd edn, Cold Spring Harbor Press, New York.

Sanda, A. I., and Ford, J.P. (1986) Genomic analysis I: inheritance units and genetic selection in the rapid discovery of locus-linked DNA markers. *Nucleic Acids Res.* **14**, 7265–7283.

Sanger, F., Nicklen, S., and Coulson, A.R. (1977) DNA sequencing with chain-terminating inhibitors. *Proc. Natl. Acad. Sci. USA* **74**, 5463–5467.

Sarkar, G., Yoon, H.-S., and Sommer, S.S. (1992a) Screening for mutations by RNA single-strand conformation polymorphism (rSSCP): comparison with DNA-SCCP. *Nucleic Acids Res.* **20**, 871–878.

Sarkar, G., Yoon, H.-S., and Sommer, S.S. (1992b) Dideoxy fingerprinting (ddF): a rapid and efficient screen for the presence of mutations. *Genomics* **13**, 441–443.

Sato, T., Tanigami, A., Yamakawa, K., Akiyama, F., Kasumi, F., Sakamoto, G., and Nakamura, Y. (1990) Allelotype of breast cancer: cumulative allele losses promote tumor progression in primary breast cancer. *Cancer Res.* **50**, 7184–7189.

Sato, T., Akiyama, F., Sakamoto, G., Kasumi, F., and Nakamura, Y. (1991) Accumulation of genetic alterations and progression of primary breast cancer. *Cancer Res.* **51**, 5794–5799.

Schäfer, R., Zischler, H., Birnser, U., Becker, A., and Epplen, J.T. (1988a) Optimized oligonucleotide probes for DNA fingerprinting. *Electrophoresis* **9**, 369–374.

Schäfer, R., Zischler, H., and Epplen, J.T. (1988b) (CAC)$_5$, a very informative oligonucleotide probe for DNA fingerprinting. *Nucleic Acids Res.* **16**, 5196.

Schlötterer, C., and Tautz, D. (1992) Slippage synthesis of simple sequence DNA. *Nucleic Acids Res.* **20**, 211–215.

Schlötterer, C., Amos, B., and Tautz, D. (1991) Conservation of polymorphic simple sequence loci in cetacean species. *Nature* **354**, 63–65.

Schuler, L.A., Weber, J.L., and Gorski, J. (1983) Polymorphism near the rat prolactin gene caused by insertion of an Alu-like element. *Nature* **305**, 159–160.

Schwartz, D.C., and Cantor, C.R. (1984) Separation of yeast chromosome-sized DNAs by pulsed field gel electrophoresis. *Cell* **37**, 67–75.

Segall, M., and Bach, F.H. (1990) HLA and disease: The perils of simplification. *N. Engl. J. Med.* **322**, 1879–1882.

Sheffield, V.C., Cox, D.R., Lerman, L.S., and Myers, R.M. (1989) Attachment of a 40–base-pair G+C-rich sequence (GC-clamp) to genomic DNA fragments by the polymerase chain reaction results in improved detection of single-base changes. *Proc. Natl. Acad. Sci. USA* **86**, 232–236.

Sheffield, V.C., Fishman, G.A., Beck, J.S., Kimura, A.E., and Stone, E.M. (1991) Identification of novel rhodopsin mutations associated with retinis pigmentosa by GC-clamped denaturing gradient gel electrophoresis. *Am. J. Hum. Genet.* **49**, 699–706.

Sheffield, V.C., Beck, J.S., Stone, E., and Myers, R.M. (1992a) A simple and efficient method for attachment of a 40–base pair, GC-rich sequence to PCR-amplified DNA. *BioTechniques* **12** (3), 386–387.

Sheffield, V.C., Beck, J.S., Nichols, B., Cousineau, A., Lidral, A.C., and Stone, E.M. (1992b) Detection of multiallele polymorphisms within gene sequences by GC-clamped denaturing gradient gel electrophoresis. *Am. J. Hum. Genet.* **50**, 567–575.

Sheffield, V.C., Beck, J.S., Kwitek, A.E., Sandstrom, D.W., and Stone, E.M. (1993) The sensitivity of single-strand conformation polymorphism analysis for the detection of single base substitutions. *Genomics* **16**, 325–332.

Shenk, T.E., Rhodes, C., Rigby, W.J., and Berg, P. (1975) Biochemical method for mapping mutational alterations in DNA with S_1 nuclease: the location of deletions and temperature- sensitive mutations in Simian virus 40. *Proc. Natl. Acad. Sci. USA* **72**, 989–993.

Sheppard, R.D., Montagutelli, X., Jean, W.C., Tsai, J.-Y., Rose, A., Guénet, J.-L., Cole, M.D., and Silver, L.M. (1991) Two-dimensional gel analysis of complex DNA families: methodology and apparatus. *Mamm. Genome* **1**, 104–111.

Shinomiya, T., and Ina, S. (1991) Analysis of chromosomal replicons in early embryos of *Drosophila melanogaster* by two-dimensional gel electrophoresis. *Nucleic Acids. Res.* **19** (14), 3935–3941.

Shore, S., and Myerowitz, R. (1990) Polymerase chain reaction generated heteroduplexes from Ashkenazi Tay–Sachs carriers with an insertion mutation can be detected on agarose gels. *Am. J. Hum. Genet.* **47**, 169.

Shuber, A.P., Skoletski, J., Stern, R., and Handelin, B.L. (1993) Efficient 12-mutation testing in the CFTR gene: a general model for complex mutation analysis. *Nucleic Acids Res.* **2**, 153–158.

Shuster, D.E., Kehrli, M.E., Jr, Ackermann, M.R., and Gilbert, R.O. (1992) Identification and prevalence of a genetic defect that causes leukocyte adhesion deficiency in Holstein cattle. *Proc. Natl. Acad. Sci. USA* **89**, 9225–9229.

Shyman, S., and Weaver, S. (1985) Chromosomal rearrangements associated with LINE elements in the mouse genome. *Nucleic Acids Res.* **13**, 5085.

Simmler, M.-C., Cox, R.D. and Avner, P. (1991) Adaptation of the interspersed repetitive sequence polymerase chain reaction to the isolation of mouse DNA probes from somatic cell hybrids on a hamster background. *Genomics* **10**, 770–778.

Simon, M., Phillips, M., and Green, H. (1991) Polymorphism due to variable number of repeats in the human involucrin gene. *Genomics* **9**, 576–580.

Sinnett, D., Deragon, J.-M., Simard, L.R., and Labuda, D. (1990) Alumorphs-human DNA polymorphisms detected by the polymerase chain reaction using Alu-specific primers. *Genomics* **7**, 331–334.

Slagboom, P.E., Mullaart, E., Droog, S., and Vijg, J. (1991) Somatic mutations and cellular aging: two-dimensional DNA typing of rat fibroblast clones. *Mutat. Res.* **256**, 311–321.

Smeets, H.J.M., Brunner, H.G., Ropers, H., and Wieringa, B. (1989) Use of variable simple sequence motifs as genetic markers: application to study of myotonic dystrophy. *Hum. Genet.* **83**, 245–251.

Smith, C.L., and Cantor, C.R. (1986) Pulsed-field gel electrophoresis of large DNA molecules. *Nature* **319**, 701–702.

Smith, C.L., Econome, J.G., Schutt, A., Klco, S., and Cantor, C.R. (1987) A physical map of the *Escherichia coli* K12 genome. *Science* **236**, 1448–1453.

Smith, F.I., Parvin, J.D., and Palese, P. (1986) Detection of single base substitutions in influenza virus RNA molecules by denaturing gradient gel electrophoresis of RNA–RNA or DNA–RNA heteroduplexes. *Virology* **150**, 55–64.

Smith, F.I., Latham, T.E, Ferrier, J.A., and Palese, P. (1988) Novel method of detecting single base substitutions in RNA molecules by differential melting behaviour in solution. *Genomics* **3**, 217–223.

Southern, E.M. (1975) Detection of specific sequences among DNA fragments separated by gel electrophoresis. *J. Mol. Biol.* **98**, 503–517.

la Spada, A.R., Wilson, E.M., Lubahn, D.B., Harding, A.E., and Fischbeck, K.H. (1991) Androgen receptor gene mutations in X-linked spinal and bulbar muscular atrophy. *Nature* **352**, 77–79.

Spandidos, D.A., and Wilkie, N.M. (1984) Malignant transformation of early passage rodent cells by a single mutated human oncogene. *Nature,* **310**, 469–475.

Spinardi, L., Mazars, R., and Theillet, C. (1991) Protocols for an improved detection of point mutations by SSCP. *Nucleic Acids Res.* **19**, 4009.

Stallings, R.L., Torney, D.C., Hildebrand, C.E., Longmire, J.L., Deaven, L.L., Jett, J.H., Doggett, N.A., and Moyzis, R.K. (1990) Physical mapping of human chromosomes by repetitive sequence fingerprinting. *Proc. Natl. Acad. Sci. USA* **87**, 6218–6222.

Steger, G., Po, T., and Riesner, D. (1987) Double-stranded cucumovirus associated RNA 5: which sequence variations may be detected by optical melting and temperature-gradient gel electrophoresis? *Nucleic Acids Res.* **15**, 5085–5103.

Steinmetz , M., Stephan, D., and Lindahl, K.F. (1986) Gene organization and recombinational hotspots in the murine Major Histocompatibility Complex. *Cell* **44**, 895–904.

Stoker, N.G., Cheah, K.S.E., Griffin, J.R., Pope, F.M., and Solomon, E. (1985) A highly polymorphic region 3′ to the human type II collagen gene. *Nucleic Acids Res.* **13**, 4613–4622.

Strezoska, Z., Paunesku, T., Radosavljevic, D., Labat, I., Dramanac, R., and Crkvenjakov, R. (1991) DNA sequencing by hybridization: 100 bases read by a non-gel-based method. *Proc. Natl. Acad. Sci. USA* **88**, 10089–10093.

Sugino, H., Oshimura, M., and Matsubara, K. (1992) Banding profiles of LTR of human endogenous retrovirus HERV-A in 24 chromosomes in somatic cell hybrids. *Genomics* **13**, 461–464.

Sulston, J., Du, Z., Thomas, K., Wilson, R., Hillier, L., Staden, R., Halloran, N., Green, P., Thierry-Mieg, J., Qiu, L., Dear, S., Coulson, A., Craxton, M., Durbin, R., Berks, M., Metzstein, M., Hawkins, T., Ainscough, R., and Waterston, R. (1992) The *C. elegans* genome sequencing project: a beginning. *Nature* **356**, 37–41.

Suzuki, Y., Orita, M., Shiraishi, M., Hayashi, K., and Sekiya, T. (1990) Detection of ras gene mutations in human lung cancers by single-strand conformation polymorphism analysis of polymerase chain reaction products. *Oncogene* **5**, 1037–1043.

Swallow, D.M., Gendler, S., Griffiths, B., Corney, G., Papadimitriou, J., and Bramwell, M.W. (1987) The human tumour-associated epithelial mucins are coded by an expressed hypervariable gene locus PUM. *Nature* **328**, 82–84.

Tabor, S., and Richardson, C.C. (1987) DNA sequence analysis with a modified bacteriophage T7 DNA polymerase. *Proc. Natl. Acad. Sci. USA* **84**, 4767–4771.

Takahashi, N., Hiyama, K., Kodaira, M., and Satoh, C. (1990) An improved method for the detection of genetic variations in DNA with denaturing gradient gel electrophoresis. *Mutat. Res.* **234**, 61–70.

T'Ang, A., Varley, J.M., Chakraborty, S., Murphree, A.L., and Fung, Y.K.T. (1988) Structural rearrangement of the retinoblastoma gene in human breast carcinoma. *Science* **242**, 263–266.

Tautz, D., and Renz, M. (1984) Simple sequences are ubiquitous repetitive components of eukaryotic genomes. *Nucleic Acids Res.* **12**, 4127–4138.

Tautz, D., Trick, M., and Dover, G.A. (1986) Cryptic simplicity in DNA is a major source of genetic variation. *Nature* **322**, 652–656.

Tautz, D. (1989) Hypervariability of simple sequences as a general source for polymorphic DNA markers. *Nucleic Acids Res.* **17**, 6463–6471.

Thein, S.L., Jeffreys, A.J., Gooi, H.C., Cotter, F., Flint, J., O'Connor, N.T.J., Weatherall, D.J., and Wainscoat, J.S. (1987) Detection of somatic changes in human cancer DNA by DNA fingerprint analysis. *Br. J. Cancer* **55**, 353–356.

Theophilus, B.D.M., Latham, T., Grabowski, G.A., and Smith, F.I. (1989) Comparison of RNase A, a chemical cleavage and GC-clamped denaturing gradient gel electrophoresis for the detection of mutations in exon 9 of the human acid-glucosidase gene. *Nucleic Acids Res.* **17**, 7707–7722.

Thibodeau, S.N., Bren, G., and Schaid, D. (1993) Microsatellite instability in cancer of the proximal colon. *Science* **260**, 816–819.

Thilly, W.G. (1985) Potential use of gradient denaturing gel electrophoresis in obtaining mutational spectra from human cells. In Huberman, E., and Barr, S.H. (eds), *Carcinogenesis*. Vol. 10. Raven, New York, pp. 511–528.

Thomson, G. (1988) HLA disease associations: models for insulin dependent diabetes mellitus and the study of complex human genetic disorders. *Annu. Rev. Genet.* **22**, 31–50.

Todd, J.A. (1992) La carte de microsatellites est arrivée! *Hum. Mol. Genet.* **1**, 9, 663–666.

Top, B., Uitterlinden, A.G., van der Zee, A., Kastelein, J.J.P., Gevers Leuven, J.A., Havekes, L.M., and Frants, R. (1992) Absence of mutations in the promotor region of the low density lipoprotein receptor gene in a large number of familial hypercholesterolemia patients as revealed by denaturing gradient gel electrophoresis. *Hum. Genet.* **89**, 561–565.

Trommelen, G.J.J.M., den Daas, J.H.G. Vijg, J., and Uitterlinden, A.G. (1993) DNA profiling of cattle using micro- and minisatellite core probes. *Anim. Genet.* **24**, 235–241.

Tsui, L.-C. (1992) The spectrum of cystic fibrosis mutations. *Trends Genet.* **8**, 392–398.

Turner, B.J., Elder, J.F., Laughlin, T.F., and Davis, W.P. (1990) Genetic variation in clonal vertebrates detected by simple sequence DNA fingerprinting. *Proc. Natl. Acad. Sci. USA* **87**, 5653–5657.

Uitterlinden, A.G., and Vijg, J. (1989) Two-dimensional DNA typing. *TIBTECH* **7**, 336–341.

Uitterlinden, A.G., and Vijg, J. (1990) Denaturing gradient gel electrophoretic analysis of the human cHa-ras1 proto-oncogene. *Appl. Theor. Electrophor.* **1**, 175–179.

Uitterlinden, A.G., and Vijg, J. (1991) Denaturing gradient gel electrophoretic analysis of minisatellite alleles. *Electrophoresis* **12**, 12–16.

Uitterlinden, A.G., Slagboom, P.E., Knook, D.L., and Vijg, J. (1989a) Two-dimensional DNA fingerprinting of human individuals. *Proc. Natl. Acad. Sci. USA* **86**, 2742–2746.

Uitterlinden, A.G., Slagboom, P.E., Johnson, T.E., and Vijg, J. (1989b) The *Caenorhabditis elegans* genome contains monomorphic minisatellites and simple sequences. *Nucleic Acids Res.* **17**, 9527–9530.

Uitterlinden, A.G., Slagboom, P.E., Mullaart, E., Meulenbelt, I., and Vijg, J. (1991a) Genome scanning by two-dimensional DNA typing: the use of repetitive DNA sequences for rapid mapping of genetic traits. *Electrophoresis* **12**, 119–134.

Uitterlinden, A.G., Mullaart, E.M., Morolli, B., and Vijg, J. (1991b) Genome scanning of higher eukaryotes by two-dimensional DNA typing using micro- and minisatellite core probes. *Methods: Companion Methods Enzymol.* **3**, 83–90.

Ullrich, A., Dull, T.J., Gray, A., Philips, J.A., and Peter, S. (1982) Variation in the sequence and modification state of the human insulin gene flanking regions. *Nucleic Acids Res.* **10**, 2225–2240.

VanWye, J.D., Bronson, E.C., and Anderson, J.N. (1991) Species-specific patterns of DNA bending and sequence. *Nucleic Acids Res.* **19** (19), 5253–5261.

Veres, G., Gibbs, R.A., Scherer, S.F., and Caskey, C.T. (1987) The molecular basis of the sparse fur mouse mutation. *Science* **237**, 415–417.

Vergnaud, G. (1989) Polymers of random short oligonucleotides detect polymorphic loci in the human genome. *Nucleic Acids Res.* **17**, 7623–7630.

Vergnaud, G., Mariat, D., Zoroastro, M., and Lauthier, V. (1991a) Detection of single and multiple polymorphic loci by synthetic tandem repeats of short oligonucleotides. *Electrophoresis* **12**, 134–140.

Vergnaud, G., Mariat, D., Apiou, F., Aurias, A., Lathrop, M., and Lauthier, V. (1991b) The use of synthetic tandem repeats to isolate new VNTR loci: cloning of a human hypermutable sequence. *Genomics* **11**, 135–144.

Verkerk, A.J.H.M., Pieretti, M., Sutcliffe, J.S., Fu, Y.-H., Kuhl, D.P.A., Pizutti, A., Reiner, O., Richards, S., Victoria, M.F., Zhang, F., Eussen, B., van Ommen, G.J.B., Blonden, L.A.J., Riggins, G., Chastain, J.L., Kunst, C.B., Galjaard, H., Caskey, C.T., Nelson, D.L., Oostra, B.A., and Warren, S.T. (1991) Identification of a gene (FMR-1) containing a CGG repeat coincident with a breakpoint cluster region exhibiting length variation in Fragile X syndrome. *Cell* **65**, 905–914.

Verwest, A.M., de Leeuw, W.J.F., Molijn, A.C., Andersen, T.I., Børresen, A.-L., Mullaart, E., Uitterlinden, A.G., and Vijg, J. (1994) Genome scanning of breast cancers by two-dimensional DNA typing. *Br. J. Cancer,* **69**, 84–92.

Vidaud, M., Vidaud, D., Siguret, V., Lavergne, J.M., and Goossens, M. (1989) Mutational insertion of an Alu sequence causes hemophilia B. *Am. J. Hum. Genet.* **45,** A226.

Vidaud, M., Fanen, P., Martin, J., Ghanem, N., Nicolas, S., and Goossens, M. (1990) Three point mutations in the CFTR gene in French cystic fibrosis patients: identification by denaturing gradient gel electrophoresis. *Hum. Genet.* **85,** 446–449.

Vijg, J., and Gossen, J. (1993) Somatic mutations and cellular ageing. *Comp. Biochem. Physiol.,* **104B,** 3, 429–437.

Vijg, J., and Uitterlinden, A.G. (1987) A search for DNA alterations in the aging mammalian genome: an experimental strategy. *Mech. Aging Dev.* **41,** 47–63.

Vijg, J., and Uitterlinden, A.G. (1991) Method for the simultaneous detection of DNA sequence variations at a large number of sites, and a kit suitable therefore. *United States Patent 5,068, 176.*

Vogelstein, B., Fearon, E.R., Hamilton, S.R., and Feinberg, A.P. (1985) Use of restriction fragment length polymorphisms to determine the clonal origin of human tumors. *Science* **227,** 642–645.

Vogt, P. (1990) Potential genetic functions of tandem repeated DNA sequence blocks in the human genome are based on a highly conserved 'chromatin folding code'. *Hum. Genet.* **84,** 301–336

Wallace, R.B., Shaffer, J., Murphy, R.F., Bonner, J., Hirose, T., and Itakura, K. (1979) Hybridization of synthetic oligodeoxyribonucleotides to φX174 DNA: the effect of single base pair mismatch. *Nucleic Acids Res.* **6,** 3543–3557.

Wallace, M.R., Anderson, L.B., Saulino, A.M., Gregory, P.E., Glover, T.W., and Collins, F.S. (1991) A *de novo* Alu insertion results in neurofibromatosis type 1. *Nature* **353,** 864–866.

Walter, M.A., and Cox, D.W. (1989) A method for two dimensional DNA electrophoresis (2D-DE): application to the immunoglobulin heavy chain variable region. *Genomics* **5,** 157–159.

Wang-Jabs, E., Goble, C.A., and Cutting, G.R. (1989) Macromolecular organization of human centromeric regions reveals high-frequency, polymorphic macro DNA repeats. *Proc. Natl. Acad. Sci. USA* **86,** 202–206.

Wartell, R.M., Hosseini, S.H., and Moran, C.P., Jr. (1990) Detecting basepair substitutions in DNA fragments by temperature-gradient gel electrophoresis. *Nucleic Acids Res.* **18,** 2699–2705.

Watson, J.D. (1990) The Human Genome Project: past, present and future. *Science* **248,** 44–49.

Watson, J.D., and Crick, F.H.C. (1953a) Molecular structure of nucleic acids: a structure for deoxyribonucleic acid. *Nature* **171,** 737–738.

Watson, J.D., and Crick, F.H.C. (1953b) Genetic implications of the structure of deoxyribonucleic acid. *Nature* **171,** 964–969.

Waye, J.S., and Fourney, R.M. (1990) Identification of complex DNA polymorphisms based on variable number of tandem repeats (VNTR) and restriction site polymorphism. *Hum. Genet.* **84,** 223–227.

Weber, J.L. (1991) Informativeness of human $(dC-dA)_n.(dG-dT)_n$ polymorphisms. *Genomics* **7,** 524–530.

Weber, J.L., and May, P.E. (1989) Abundant class of human DNA polymorphisms which can be typed using the polymerase chain reaction. *Am. J. Hum. Genet.* **44**, 388–396.

Weissenbach, J., Gyapay, G., Dib, C., Vignal, A., Morisette, J., Millasseau, P., Vaysseix, G., and Lathrop, M. (1992) A second-generation linkage map of the human genome. *Nature* **359**, 794–801.

Welsh, J., and McClelland, M. (1990) Fingerprinting genomes using PCR with arbitrary primers. *Nucleic Acids Res.* **18**, 7213–7218.

Welsh, J., Petersen, C., and McClelland, M. (1991) Polymorphisms generated by arbitrarily primed PCR in the mouse: application to strain identification and genetic mapping. *Nucleic Acids Res.* **19**, 303–306.

Weston, A., Willey, J.C., Modali, R., Sugimura, H., McDowell, E.M., Resau, J., Light, B., Haugen, A., Mann, D.L., Trump, B.F., and Harris, C.C. (1989) Differential DNA sequence deletions from chromosomes 3, 11, 13, and 17 in squamous-cell carcinoma, large-cell carcinoma, and adenocarcinoma of the human lung. *Proc. Natl. Acad. Sci. USA* **86**, 5099–5103.

Wetmur, J.G., Ruyechan, W.T., and Douthart, R.J. (1981) Denaturation and renaturation of penicillin chrysogenum mycophage double-stranded ribonucleic acid in tetraalklammonium salt solutions. *Biochemistry* **20**, 2999–3002.

White, M.B., Krueger, L.F., Holsclaw, D.S., Gerrard, B., Stewart, C., Quittel, L., Dolganov, G., Baranov, V., Ivaschenko, T., Kapranov, N.I., Sebastio, G., Castiglione O., and Dean, M. (1991) Detection of three rare frameshift mutations in the cystic fibrosis gene in an African American (CF444delA), an Italian (CF2522insC), and a Soviet (CF3821delT). *Genomics* **10**, 266–269.

White, M.-B, Carvalho, M., Derse, D., O'Brien, S.J., and Dean, M. (1992) Detecting single base substitutions as heteroduplex polymorphisms. *Genomics* **12**, 301–306.

White, R., Lippert, M., Bishop, D.T., Barker, D., Berkowitz, J., Brown, C., Callahan, P., Holmes, T., and Jerominsky, L. (1985) Construction of linkage maps with DNA markers for human chromosomes. *Nature* **313**, 101–105.

Wiggs, J., Nordenskjöld, M., Yandell, D., Rapaport, J., Grondin, V., Janson, M., Werelius, B., Peterson, R., Craft, A., Riedel, K., Lieberfarb, R., Walton, D., Wilson, W., and Drya, T. (1988) Prediction of the risk of hereditary retinoblastoma, using DNA polymorphisms within the retinoblastoma gene. *New Engl. J. Med.* **318**, 151–157.

Willard, H.F., Waye, J.S., Skolnick, M.H., Schwartz, C.E., Powers, V.E., and England, S.B. (1986) Detection of restriction fragment length polymorphisms at the centromeres of human chromosomes by using chromosome-specific α-satellite DNA probes: implications for the development of centromere-based linkage maps. *Proc. Natl. Acad. Sci. USA* **83**, 5611–5615.

Williams, J.G.K., Kubelik, A.R., Livak, K.J., Rafalski, J.A., and Tingey, S.V. (1990) DNA polymorphisms amplified by arbitrary primers are useful as genetic markers. *Nucleic Acids Res.* **18**, 6531–6535.

Winter, E.E., Yamamoto, E., Almoguera, C., and Perucho, M. (1985) A method to detect and characterize point mutations in transcribed genes: amplification and over-expression of the mutant c-Ki-ras allele in tumor cells. *Proc. Natl. Acad. Sci. USA* **82**, 7575–7579.

Wolff, R.K., Plaetke, R., Jeffreys, A.J., and White, R. (1989) Unequal crossing over between homologous chromosomes is not the major mechanism involved in the generation of new alleles at VNTR loci. *Genomics* **5**, 382–384.

Womack, J.E. (1992) Molecular genetics arrives on the farm. *Nature* **360**, 108–109.

Wong, Z., Wilson, V., Jeffreys, A.J., and Thein, S.L. (1986) Cloning of a selected fragment from a human DNA fingerprint: isolation of an extremely polymorphic minisatellite. *Nucleic Acids Res.* **14**, 4605–4616.

Wong, Z., Wilson, V., Patel, I., Povey, S., and Jeffreys, A.J. (1987) Characterization of a panel of highly variable minisatellites cloned from human DNA. *Ann. Hum. Genet.* **51**, 269–288.

Woods-Samuels, P., Wong, C., Mathias, S.L., Scott, A.F., Kazazian, H.H., and Antonorakis, S.E. (1989) Characterization of a non-deleterious L1 insertion in an intron of the human factor VIII gene and further evidence of open reading frames in functional L1 elements. *Genomics* **4**, 290–296.

Woolf, T., Lai, E., Kronenberg, M., and Hood, L. (1988) Mapping genomic organization by field inversion and two-dimensional gel electrophoresis: application to the murine T-cell receptor τ gene family. *Nucleic Acids Res.* **16** (9), 3863–3875.

Wyman, A.R., and White, R. (1980) A highly polymorphic locus in human DNA. *Proc. Natl. Acad. Sci. USA* **77**, 6754–6758.

Yandell, D.W., and Dryja, T.P. (1989) Detection of DNA sequence polymorphisms by enzymatic amplification and direct genomic sequencing. *Am. J. Hum. Genet.* **45**, 547–555.

Yee, T., and Inouye, M. (1982) Two-dimensional DNA electrophoresis applied to the study of DNA methylation and the analysis of genome size in *Myxococcus xanthys*. *J. Mol. Biol.* **154**, 181–196.

Yee, T., and Inouye, M. (1983) Two-dimensional DNA electrophoresis methods utilizing *in situ* enzymatic digestions. In: *Experimental manipulation of gene expression* (M. Inouye, Ed.), Academic Press, New York, USA, pp. 279–290.

Yee, T., and Inouye, M. (1984) Two-dimensional S1 nuclease heteroduplex mapping: detection of rearrangements in bacterial genomes. *Proc. Natl. Acad. Sci. USA* **81**, 2723–2727.

Yi, M., Au, L.-C., Ichikawa, N., and Ts'o, P.O.P. (1990) Enhanced resolution of DNA restriction fragments: a procedure by two-dimensional electrophoresis and double-labelling. *Proc. Natl. Acad. Sci. USA* **87**, 3919–3923.

Yu, S., Pritchard, M., Kremer, E., Lynch, M., Nancarrow, J., Baker, E., Holman, K., Mulley, J.C., Warren, S.T., Schlessinger, D., Sutherland, G.R., and Richards, R.I. (1991) Fragile X genotype characterized by an unstable region of DNA. *Science* **252**, 1179–1181.

Zechner, R., Newman, T.C., Steiner, E., and Breslow, J.L. (1991) The structure of the mouse lipoprotein lipase gene: B1 repetitive element is inserted into the 3′ untranslated region of the mRNA. *Genomics* **11**, 62–76.

Zhuang, J., Constantinou, C.D., Ganguly, A., and Prockop, D.J. (1991) A single base mutation in type I procollagen (COLIAI) that converts glycine αI-541 to aspartate in a lethal variant of osteogenesis imperfecta: detection of the mutation with a carbodiimide reaction of DNA heteroduplexes and direct sequencing of products of the PCR. *Am. J. Hum. Genet.* **48**, 1186–1191.

Zietkiewicz, E., Labuda, M., Sinnett, D., Glorieux, F.H., and Labuda, D. (1992) Linkage mapping by simultaneous screening of multiple polymorphic loci using Alu oligonucleotide-directed PCR. *Proc. Natl. Acad. Sci. USA* **89**, 8448–8451.

Zischler, H., Nanda, I., Schäfer, R., Schmid, M., and Epplen, J.T. (1989) Digoxigenated oligonucleotide probes specific for simple repeats in DNA fingerprinting and hybridization *in situ. Hum. Genet.* **82**, 227–233.

Zuliani, G., and Hobbs, H.H. (1990) A high frequency of length polymorphisms in repeated sequences adjacent to Alu sequences. *Am. J. Hum. Genet.* **46**, 963–969.

Index